DESIGNING HEALTHCARE
THAT WORKS

DESIGNING HEALTHCARE THAT WORKS

A SOCIO-TECHNICAL APPROACH

Edited by

MARK S. ACKERMAN

SEAN P. GOGGINS

THOMAS HERRMANN

MICHAEL PRILLA

CHRISTIAN STARY

ACADEMIC PRESS

An imprint of Elsevier

Academic Press is an imprint of Elsevier
125 London Wall, London EC2Y 5AS, United Kingdom
525 B Street, Suite 1800, San Diego, CA 92101-4495, United States
50 Hampshire Street, 5th Floor, Cambridge, MA 02139, United States
The Boulevard, Langford Lane, Kidlington, Oxford OX5 1GB, United Kingdom

Library of Congress Cataloging-in-Publication Data
A catalog record for this book is available from the Library of Congress

British Library Cataloguing-in-Publication Data
A catalogue record for this book is available from the British Library

ISBN: 978-0-12-812583-0

For information on all Academic Press publications visit our website at
https://www.elsevier.com/books-and-journals

Working together
to grow libraries in
developing countries

www.elsevier.com • www.bookaid.org

Publisher: Mica Haley
Acquisition Editor: Rafael E. Teixeira
Editorial Project Manager: Fenton Coulthurst
Production Project Manager: Anusha Sambamoorthy
Cover Designer: Christian J. Bilbow

Typeset by TNQ Books and Journals

Contents

List of Contributors

Khuloud Abou Amsha Troyes University of Technology, Troyes, France

Mark S. Ackerman University of Michigan, Ann Arbor, MI, United States

Martina Augl Tirol Kliniken, Innsbruck, Austria

Jakob E. Bardram Technical University of Denmark, Lyngby, Denmark

Claus Bossen Aarhus University, Aarhus, Denmark

Randall S. Burd Children's National Medical Center, Washington, DC, United States

Ayşe G. Büyüktür University of Michigan, Ann Arbor, MI, United States

Carmen Castaneda-Sceppa Northeastern University, Boston, MA, United States

Mads M. Frost IT University of Copenhagen, Copenhagen, Denmark

Sean P. Goggins University of Missouri, Columbia, MO, United States

Thomas Herrmann Ruhr University of Bochum, Bochum, Germany

Jessica A. Hoffman Northeastern University, Boston, MA, United States

Pei-Yao Hung University of Michigan, Ann Arbor, MI, United States

Maia Jacobs Georgia Institute of Technology, Atlanta, GA, United States

Myriam Lewkowicz Troyes University of Technology, Troyes, France

Ivan Marsic Rutgers University, Piscataway, NJ, United States

Michelle A. Meade University of Michigan, Ann Arbor, MI, United States

Elizabeth D. Mynatt Georgia Institute of Technology, Atlanta, GA, United States

Mark W. Newman University of Michigan, Ann Arbor, MI, United States

Andrea G. Parker Northeastern University, Boston, MA, United States

Michael Prilla Clausthal University of Technology, Clausthal-Zellerfeld, Germany

Herman Saksono Northeastern University, Boston, MA, United States

Aleksandra Sarcevic Drexel University, Philadelphia, PA, United States

Christian Stary University of Linz, Linz, Austria

Kai Zheng University of California – Irvine, Irvine, CA, United States

Xiaomu Zhou Northeastern University, Boston, MA, United States

Biography of Authors

Khuloud Abou Amsha is a postdoctoral fellow at Troyes University of Technology (France). Her research area is Human-Computer Interaction (HCI), primarily in Computer-Supported Cooperative Work (CSCW). Her research focuses on analyzing and designing applications that support collaborative activities. She is interested in new ways of organizing collaborations that go beyond the boundaries of organizations. She was involved in several projects in the field of health care at the regional and national levels.

Mark S. Ackerman is the George Herbert Mead Collegiate Professor of Human–Computer Interaction and a Professor in the School of Information and in the Department of Electrical Engineering and Computer Science at the University of Michigan, Ann Arbor. He also has a courtesy appointment in the Department of Learning Health Sciences in the College of Medicine. Mark's major research area is human–computer interaction (HCI), primarily social computing and computer-supported cooperative work (CSCW). Mark has published widely in HCI/CSCW, including expertise sharing (especially for chronic conditions), collaborative information access, and pervasive environments for health. His work spans both technical and social analytic studies. For his work on socio-technical systems, Mark was elected as a member of the CHI Academy and as an ACM Fellow. Previously, Mark was faculty at the University of California, Irvine, and a research scientist at MIT's Laboratory for Computer Science (now CSAIL).

Martina Augl is head of the Organizational Development Team at Tirol Kliniken GmbH, a leading regional health care provider in Austria. She designs and implements organizational change processes utilizing knowledge management techniques, to improve the effectiveness and agility of organizational units. Promoting a stakeholder-centered approach she has managed to streamline different organizational cultures toward focused patient care and quality-assured health care procedures.

Jakob E. Bardram is a professor at the Technical University of Denmark, where he directs the Copenhagen Center for Health Technology (www.cachet.dk). His research interests include pervasive computing, HCI, and software architecture, with a special focus on health technologies. Bardram has a PhD in computer science from the University of Aarhus, Denmark.

Claus Bossen conducts research into health care and health care IT, including the design, development, implementation, and evaluation of such technologies. Especially creating a fit between technology, work practices and organization to ensure that technologies support and augment work rather than hinder and make is more troublesome has been at the center for attention, and he has conducted extensive research into these issue around electronic health care records. Most recently he has started research into health care data and governance including value-based health care and diagnose-related groups. With a PhD in anthropology he is especially interested in qualitative methods, including interviews, participant-observation, and more generally research that entails engagements with people. Research fields include anthropology, participatory design, computer-supported cooperative work, and health care informatics. He is presently associate professor at Aarhus University, Denmark.

Randall S. Burd is a pediatric surgeon and the Chief of the Division of Trauma and Burn Surgery, at Children's National Medical Center, Washington, DC, USA. He is also a Professor of Surgery and Pediatrics at the George Washington School of Medicine and Health Sciences. His research focuses on developing novel strategies for improving the early care of injured children. Dr. Burd's research has been funded by the US National Institutes of Health (NIH), US Department of Health and Human Services (DHHS), and US National Science Foundation (NSF). Dr. Burd received the BA degree from Dartmouth University in 1983, the MD degree from Columbia University in 1987, and the PhD degree from the University of Minnesota in 1994, where he also completed his residency in surgery. He is trained in pediatric surgery at the Children's Hospital of Michigan in Detroit, MI.

Ayşe G. Büyüktür is a researcher at the University of Michigan School of Information. She is also affiliated with the Michigan Institute for Clinical and Health Research. Her primary research interests are at the intersections of health informatics, computer-supported cooperative work, and human–computer interaction. She is particularly interested in the design and use of health information technologies to manage and coordinate care in chronic illness and disability. Ayşe received her PhD in Information, and her Master's degrees in Public Health and Pharmaceutical Sciences, from the University of Michigan.

Carmen Castaneda-Sceppa is professor and Chair of the Department of Health Sciences in the Bouve College of Health Sciences at Northeastern University. She holds an MD from Francisco Marroquin University, Guatemala, and a PhD in Nutrition from Tufts University. Dr. Sceppa's translational research focuses on the design, evaluation, and dissemination of nutrition and physical activity/exercise interventions that promote health and reduce the risk of chronic conditions among older adults and through the lifespan.

Mads M. Frost is a post doc at the IT University of Copenhagen and a member of ITU's Pervasive Interaction Technology Lab. His research interests include pervasive computing and HCI, with a special focus on personal health technologies. Frost has a PhD in computer science from ITU.

Sean P. Goggins is an associate professor at Missouri's iSchool, with courtesy appointments as core faculty in the University of Missouri Informatics Institute and Department of Computer Science. He teaches, publishes, and conducts research on the uptake and use of information and communication technologies by small groups in medium to large-scale socio-technical systems; from Facebook, to online course systems and open online communities. Sean conceptualizes "group informatics" as a methodological approach and ontology for making sense of the interactions between people in medium to large-scale social computing environments. His research examines the information behavior, knowledge construction, identity development, performance, and the structural evolution of small, online groups. Goggins' recent work focuses on large-scale meta studies of open online communities, measuring performance in virtual organizations, and identifying factors that influence performance and learning in technology-mediated environments. Goggins is currently leading the development of gaming analytics on the Mission Hydro Sci projects at the University of Missouri.

Thomas Herrmann is a professor of Information and Technology-Management at the Institute of Applied Work Science (IAW), University of Bochum, Germany since 2004, and a fellow of the Electrical Engineering Department. Current research interests include design methods for socio-technical systems in various areas such as health care, computer-supported collaboration, knowledge management, (work-)process management. He teaches courses in groupware, knowledge management, socio-technical systems design, information systems and privacy, human–computer interaction, organizational communication, and process management. He has developed, evaluated, and refined a method which combines socio-technical design and process management and works together with several institutions in the health care sector. He has specialized on the design of socio-technical processes and has published more than 100 papers in this area.

He was an Associate Professor from 1992 to 2004 at the Computer Science Department at the University of Dortmund and was in charge of the development of infrastructure and new media for the University of Dortmund as a vice president from 2002 to 2004. Currently he is the privacy protection officer of the University of Bochum and a member of Paluno—The Ruhr Institute for Software Technology.

Jessica A. Hoffman is an associate professor at Northeastern University in the Department of Applied Psychology. She holds a PhD in School Psychology and an MEd in Human Development from Lehigh University and a BA in Psychology from Hamilton College. Dr. Hoffman's research focuses on the design, implementation, and evaluation of interventions that promote healthy eating and physical activity among children.

Maia Jacobs is a PhD candidate in Georgia Institute of Technology's Human-Centered Computing department. Her primary research focus is in ubiquitous computing and personal health informatics, developing and evaluating tools to support individuals' dynamic personal health management needs and goals. In her work, Maia utilizes qualitative research methods and longitudinal technology deployments to understand the benefits and limitations for personalized and adaptive support tools for individuals managing chronic illnesses. This research has gained national attention, having recently been recognized in the 2016 report to the President of the United States from the President's Cancer Panel, which focuses on improving cancer-related outcomes. In 2018, Maia will join Harvard University as a CRCS Postdoctoral Fellow.

Myriam Lewkowicz is full professor in Informatics at Troyes University of Technology (France), where she is head of Information System Management teaching branch and Tech-CICO research group (part of UMR CNRS 6281). Her interdisciplinary research consists in defining digital technology to support existing collective practices or to design new collective activities. For 10 years now, her main application domain is health care, with a focus on fostering social support among people in difficult situations (informal caregivers, isolated elderly people), and another one on supporting coordination among professionals.

Ivan Marsic is a Professor of Electrical and Computer Engineering in the School of Engineering at Rutgers University, New Jersey, USA. His research interests are in software engineering and sensor networks for health care applications, advanced groupware systems to support distributed multimedia collaboration, and gesture- and speech-based human -computer interfaces. Dr. Marsic's research has been funded by the US National Science Foundation (NSF), US National Institutes of Health (NIH), US Army, and several corporate sponsors. He received the BS and MS degrees in Computer Engineering from the University of Zagreb, Croatia, and the PhD degree in Biomedical Engineering from Rutgers University in 1994.

Michelle A. Meade is an associate professor in the Department of Physical Medicine and Rehabilitation in the School of Medicine at the University of Michigan and Clinical Rehabilitation Psychologist in Michigan Medicine. She provides psychotherapy for and conducts research with adolescents and adults with spinal cord disorders and other physical disabilities. Her primary focus is on optimizing health and quality of life, promoting self-management, and reducing health care disparities among individuals with disabilities.

Elizabeth D. Mynatt is the executive director of Georgia Tech's Institute for People and Technology, a distinguished professor of interactive computing, and the director of the everyday computing lab. She investigates the design and evaluation of health information technologies across multiple chronic care conditions including cancer, diabetes, epilepsy, and mental health. Mynatt is also the Chair of the Computing Community Consortium, an NSF-sponsored effort to engage the computing research community in envisioning more audacious research challenges. She serves as member of the National Academies Computer Science and Telecommunications Board (CSTB) and as an ACM Council Member at Large. She has been recognized as an ACM Fellow, a member of the SIGCHI Academy, and a Sloan and Kavli research fellow.

Mark W. Newman is an associate professor in the School of Information at the University of Michigan, Ann Arbor. His research interests lie broadly in the field of human-computer interaction, and particularly in the areas of ubiquitous computing and end-user programming. Mark's research group is the Interaction Ecologies group. Before joining UMSI, Mark was a research scientist at the Palo Alto Research Center (PARC, formerly known as Xerox PARC) and a doctoral candidate in Computer Science at UC Berkeley.

Andrea G. Parker is an assistant professor at Northeastern University, with joint appointments in the College of Computer & Information Science and the Bouvé College of Health Sciences. She holds a PhD in human-centered computing from Georgia Tech and a B.S. in computer science from Northeastern University. Dr. Parker directs the Wellness Technology Lab at Northeastern University, where she designs, implements, and evaluates technological approaches to reducing health disparities. Her research contributes to the fields of human–computer interaction, computer-supported cooperative work, and personal health informatics.

Michael Prilla is a professor for Human-Centered Information Systems at Clausthal University of Technology, Clausthal-Zellerfeld, Germany. He has written more than 100 research papers and book chapters. His research interests are cooperation, cooperative work in mixed and augmented reality, learning, and reflection at work, socio-technical systems and IT in health care environments, ubiquitous computing and the convergence of digital and physical interaction. He has published in the Communications of the ACM and in conferences such as ACM Group, ECIS, and ECSCW. He is deputy chair of the German special interest group on CSCW and a member of the steering committee of the German SIG on human–computer interaction. He serves on committees for conferences like ACM Group, CSCW, ECSCW, and CHI.

Herman Saksono is a computer science PhD student at Northeastern University. He holds a Master's degree in computer science from Northeastern University and a B.Eng. in Electrical Engineering from Universitas Gadjah Mada in Indonesia. Herman's research examines how technologies can be designed to positively influence human behavior in the context of health promotion and community empowerment.

Aleksandra Sarcevic is an assistant professor of Information Science in the College of Computing and Informatics at Drexel University, Pennsylvania, USA, where she directs the Interactive Systems for Healthcare (IS4H) Research Lab. Her research interests are in computer-supported cooperative work and medical informatics, with a focus on ethnographic studies of practice in safety-critical medical settings to inform technology design and implementation. Dr. Sarcevic's research is supported by the US National Institutes of Health (NIH) and US National Science Foundation (NSF). She was awarded a 2013 National Science Foundation Early CAREER Grant to support her work on information technology design and development for fast-response medical teams. Dr. Sarcevic received the MS and PhD degrees from the School of Communication and Information, Rutgers University, in 2005 and 2009, respectively.

Christian Stary is currently full professor and Chair of the Department of Business Information Systems—Communications Engineering as well as Head of the Knowledge Management Competence Center at the University of Linz, Austria. He researches the design of learning support systems, using knowledge-based and organizational development techniques. He initiates and manages projects targeting intelligent design while focusing on the needs and capabilities of stakeholders. He has been involved in several national health care projects with respect to socio-technical developments of clinics' organization of work. Thereby, modeling and process management plays a crucial role, besides epistemologically grounded method development, in particular aligning value network analysis and subject-oriented business process management. His work is reflected in a number of scientific journal contributions, international conferences, and workshops on cross-disciplinary systems engineering research. He serves in several international scientific bodies with key responsibilities, such as Editor-in-chief of Springer's Journal of Interaction Science, and Board Chair of the International Council on Knowledge Management.

Kai Zheng is associate professor of Informatics and Associate Adjunct Professor of Emergency Medicine at the University of California, Irvine (UCI). He is also co-director of the Center for Biomedical Informatics at the UCI Institute for Clinical and Translational Science. Zheng's research draws upon techniques from the fields of information systems and human–computer interaction to study the use of information, communication, and decision technologies in patient care delivery and management. His recent work has focused on topics such as interaction design, workflow and socio-technical integration, and diffusion and evaluation of health IT. Zheng received his PhD degree in Information Systems from Carnegie Mellon University. He is the recipient of the 2011 American Medical Informatics Association New Investigator Award that recognizes early informatics contributions and significant scholarly achievements.

Xiaomu Zhou is an assistant teaching professor at Northeastern University, where she leads the Master of Science in Informatics Program in the College of Professional Studies. Her teaching and research focus on human–computer interaction (HCI), health informatics, and computer-supported cooperative work (CSCW). Her managerial responsibilities include new curriculum development, faculty mentoring, and building business partnerships, with the ultimate goal of improving education. Prior to joining Northeastern, Dr. Zhou worked as an assistant professor in the School of Communication and Information at Rutgers University. She received her PhD from the School of Information at University of Michigan, Ann Arbor. She co-authored a number of papers focusing on understanding clinical documentation processes, patient-provider communication, and chronic illness management from the perspectives of diabetic and depressed patient populations.

Introduction

What if doctors, nurses, health administrators, and even (non-)patients become capable to design health care systems? A fantasy? We do not think so. When closer looking to the increased stakeholder involvement in system developments in constituting elements of societies, such as education, economy, and health care, it rather appears to become a necessity. Why is that? Particularly, health care systems have become increasingly complex, not only because of increasing in-depth expertise and the subsequent diversification of medical fields but also because of cost pressure, volatile settings, external changes (e.g., social security, insurance reform), increased speed of scientific research and innovation, and last, but not least, high expectations with respect to patient care and life quality for the elderly.

This book presents and analyses IT support for different aspects of health care in the light of socio-technical systems, in which technical developments are integrally connected to social dynamics and the needs of users. The different cases studies in this book illustrate how technology and social systems can be analyzed, developed, and sustained together.

This perspective is especially helpful for health care: supporting a system and helping it to operate despite this quickly increasing complexity can be achieved by revisiting the nature of the involved systems—in particular technology, people, and their interface— viewing them as socio-technical systems. This is a key perspective for IT systems in health care. For example, whenever a citizen gets in touch with a health care system, personal data are processed and needed for further activities, ranging from diagnosis to treatment planning, treatment, and payment. At the same time, various professional user groups, such as medical experts, care takers, administration, political bodies, must get in touch with technical systems when addressing health care issues.

Originally, socio-technical design referred to organizational redesign based on a specific theoretical basis and a strong methodology (e.g., Mumford, 1987–2003). It argues "that when new work systems are being designed equal weight should be given to social and technical factors. It places great emphasis on improving the quality of working life" rather than on gaining competitive advantage (Mumford, 1994, p. 313). In this volume, we enrich the definition of "socio-technical design" to include as central foci not only of organizational issues but also the technology itself, social interactions and dynamics outside organizations, people's practices, and so on. This is common usage in human–computer interaction, computer-supported cooperative work, science and technology studies, and software engineering, among other areas. This enrichment establishes a practical theory for health care domains.

Who should get involved when tackling socio-technical design issue and becoming aware of opportunities, barriers, and capabilities? Anyone can become an active designer concerned with and involved in the design and implementation of health care systems, including software designers, medical informatics professionals, health care administration, and potential patients. This book provides several examples in which actors often not involved into the design of IT in health care were key to the success of the socio-technical design process. In particular, practitioners and academics from areas such as information systems, public health, health and medical informatics, and user experience/human–computer interaction need to become knowledgeable and skilled in health care design processes to create usable, useful, and sustainable solutions for all stakeholders. Finally, and maybe most important, we need to give the stakeholders in the day-to-day working of health care (doctors, nurses etc.) agency to make changes in the design of their socio-technical systems.

In this volume, we go beyond traditional software design and implementation for health care, as we take up the pragmatic, messy problems of designing and implementing socio-technical solutions. They integrate organizational and technical systems for the benefit of human health. There are a broad range of individual and social (organizational, institutional, societal) needs that must be addressed. Since health care IT systems are notoriously expensive, difficult to implement, and hard to manage for health care providers, we believe the cases presented in this volume will be valuable and should be broadly considered by people responsible for and affected by the design and implementation of health care IT systems.

With the presented cases, we intend to help practitioners to apply principles of socio-technical design in health care, and consider the adoption of new theories of change. Practitioners need new processes and tools to create a more systematic alignment between technical mechanisms and social structures in health care.

To make health care IT more closely suited to care provision, this systematic alignment of the social and technical will, when appropriate, include consideration of organizational change. For patient-facing applications, social issues more broadly construed will be important. We must also recognize that the requirements of the alignment between social and technical aspects in health care are dynamic. For example, the inertia and local dynamics in health care often present substantive obstacles to the implementation of socially and technically aligned systems. Practitioners therefore need more adaptive techniques for evaluating progress and measuring system impacts.

The systematic understanding developed within this book's chapters includes new ways of designing and adopting socio-technical systems in health care. For example, as often mentioned in the literature, this might include helping practitioners examine the role of exogenous factors like health-related quality of life. Or, more globally, helping practitioners learn how to consider systems external to the boundaries drawn around a particular health care IT system is another key design challenge.

To serve designers and managers of a broad range of health care IT systems, each case study focuses on specific projects and covers an iterative cycle of socio-technical design in health care (and in some cases, its entire lifecycle). Each case represents an empirical, field-based study situated in a health care organization, network, or community. To deepen the understanding of the socio-technical challenges in health care design, each case reflects on the social and technical obstacles designers need to overcome. We complement North American health care views on socio-technical design with a European perspective.

Qualitative descriptive reporting from field studies has turned out useful for demonstrating the practicability of (novel) paradigms and the state of affairs in several fields. Each study in this book therefore describes socio-technical design focused on an essential issue, such as qualifying caregivers for planning daily patient routines. This does not only include a description or story, but also an analysis, which concerns the design problem that has been solved. Each solution is described in terms of a socio-technical design approach, the results, detailing design accomplishments, and the design process.

Finally, an evaluation for each solution is described, before summarizing the lessons learned from that case. The case studies can be considered from various eHealth aspects related to stakeholder support as indicated in the table:

- Application domain/field of intervention: this category refers to the addressed health care (IT) domain or field of concerned socio-technical design activities in the case.
- System architecture/tool chain/(organizational) IT device or application: this perspective provides insights into the technical system (including its structure) that has been used, designed, and/or provided, however, with respect to organizational effects or capabilities. It also refers to integrated infrastructures or platforms built to provide health care services.
- Methodological approach: this category comprises all methodological details, either with respect to existing concepts and methods, or innovative approaches and formats.
- Achievements according to objectives: last but not least, the effect of the case intervention is provided with respect to its objectives.

Table 1 provides an overview of the chapters.

In their chapter, Mark S. Ackerman, Ayşe G. Büyüktür, Pei-Yao Hung, Michelle A. Meade, and Mark Newman describe the SCILLS (the Spinal Cord Injury Living and Learning System), a system to help spinal cord injury (SCI) patients to acquire self-care skills and develop

TABLE 1   Designing Health care for Stakeholder Support

Chapter	Application Domain/ Field of Intervention	System Architecture/Tool Chain/(Organizational) IT Device or Application	Methodological Approach	Achievements According to Objectives
Ackerman et al.	Home care and remote monitoring of patients	Sensor network, patient tracking, and alerting system (SCILLS), clinician alerting (SCILLS)	Human-centered design process, interviews, technical probe	Understanding of sensor networks and limitations for supporting disabled
Jacobs and Mynatt	Personalized cancer treatment—guidelines for socio-technical design	Tablet app as navigator for nurses to personalize long-term cancer care	Interview Focus group Technology probing after navigation course study (journey compass)	Improving cancer care in daily practice of life
Abou Amsha and Lewkowicz	Home care of patients	Liaison notebook (CARE local tablet app) as coordinative artifact—large degree of freedom of use	Design case study Ethnography Scenario-based design	Coordination improvement Timely information of patients w.r.t. activities or treatment

(Continued)

TABLE 1   Designing Health care for Stakeholder Support—cont'd

Chapter	Application Domain/ Field of Intervention	System Architecture/Tool Chain/(Organizational) IT Device or Application	Methodological Approach	Achievements According to Objectives
Parker et al.	Personal health care support in community-based organizations, such as schools	Pervasive game app (stationary in community center) with self-monitoring activity trackers for Spaceship launch	Design case study Focus group Gamification	Clarification of cost/ownership of intervention/ process structure, sharing experiences, increasing collective activity
Bossen	Models of action: linear, rationalistic, interactional	Self-organization of work: employees themselves coordinate and ensure quality and efficiency	Focus on problems from an intrinsically social and technical perspective at the same time; principle of equi-finality	Incorporation of learning and negotiation is incorporated in continuous design process
Augl and Stary	Workforce planning	Subject-oriented web-based planning support, communication-based interaction support	Articulation of planning practice (interactions)— process redesign—process execution— evaluation	Improve daily clinical workforce planning to take better care about patients → improved availability
Sarcevic et al.	Real-time information transparency between doctor and patients	Real-time dashboard (TRUBoard) according to information needs	Observation Interview Simulation/ Focus group/ Design workshops	Supporting real-time information capturing and delivering while doctors interact with patients
Zhou et al.	Use of electronic health record	eCare electronic health record	Ethnography	Better representation of patient over the long term
Prilla and Herrmann	Patient relatives'— doctor interaction	Tablet to collect reflection data, Web-based access	Ethnographic study/ prototyping— formative evaluation	Improve socio-technical setting in interaction between doctors and patient relatives
Bardram and Frost	Enriching perspective on development through stakeholder (group) recognition	Mobile app for self-diagnosis and psychiatrist feedback when looking at the input data (personal health care apps)	Design case study Ethnography Scenario-based design	Cost reduction through personal health care technology supporting patients

self-care plans. It does this within a "smart" sensor-rich environment that helps monitor people with SCI and provides feedback to them, their caregivers, and clinicians. In this chapter, using a human-centered design process, the authors describe the basic design of the system, the requirements analysis and formative evaluation (which analyzes some of the basic practices that people with SCI perform), and the technical lessons learned. They note there were significant issues with trying to incorporate a sensor-based environment; they point to what they call the issues of sensor completeness and computational completeness in the technical design. The authors close with an analysis of what it means to design within a rapidly changing technical environment, one where the capabilities and constraints are unclear. The problem, as uncovered in this work, is understanding how the social requirements and the unclear technical will join together over time, which the authors call the socio-technical trajectory problem in design.

Maria Jacobs and Elizabeth D. Mynatt develop design principles for supporting patient-centered journeys addressing chronic illnesses. Since greater responsibilities are placed on patients to manage their health while away from traditional health care settings, personal health management tools need to catch up helping individuals with their health management. Rather than focusing on isolated tasks or events, they need to be able to tackle a broad range of needs as patients grapple with physical, emotional, and logistical challenges throughout care. Further, these challenges change as patients' needs and goals shift over time; thus, these illnesses are better defined as a dynamic journey than a series of singular events. The authors have assessed how personal mobile technologies, integrated into a health care delivery system, may better support an individual's health care journey, using breast cancer as a case study. The case study includes examining the practices of cancer navigators, characterizing how survivors describe their cancer journeys, and conducting a pilot study of tablet computers designed to offer holistic support to newly diagnosed breast cancer patients. Utilizing observations from this case study, a set of design guidelines can be offered for supporting patients' personal health management while considering the broad range of challenges that comprise this work.

In their chapter "Supporting Collaboration to Preserve the Quality of Life of Patients at Home—A Design Case Study", Khuloud Abou Amsha and Myriam Lewkowicz, present a study on introducing an innovative way of organizing home care in the city of Troyes (in the northeast of France). They have observed the collaborative practices of a group of self-employed care professionals. Among this group, collaboration occurs in episodes depending on the requirement of the patient's situation. The chapter identifies: (1) the centrality of coordinative artifacts; (2) the complexity of addressing issues beyond the medical scope; (3) the adoption of different rhythms of collaboration depending on the patient's situation. These findings led the authors to define some implications for design, that were discussed during design workshops, and that they implemented in the CARE application. The authors observed the use of CARE at five patients' homes during 5 months. This pilot study helped the authors to identify three topics of importance for supporting collaboration in home care context: (1) ensuring flexibility to accommodate different values, (2) building trust, and (3) open sharing.

Andrea G. Parker, Herman Saksiono, Jessica A. Hoffman, and Carmen Castaneda-Sceppa contribute to community health orientation for wellness technology design and delivery by addressing health promotion increasingly occurring outside of the boundaries of traditional care settings such as hospitals and clinics. They investigated interventions that are anchored within community-based organizations, seeking to meet the nuanced needs of local residents as a vital component of the wellness promotion ecosystem. These programs are particularly critical when addressing health in low-socioeconomic communities, since the services need to be more accessible, affordable, and relevant to the needs of populations facing significant barriers to wellness. The authors examine how technology can become embedded in the context of community-based health interventions. In their research, they employed a user-centered design process to create and evaluate a novel family *exergame* within a community-based organization. It allows discussing the criticality, challenge, and benefits of integrating wellness technologies within a broader community health promotion infrastructure.

Claus Bossen builds on an underlying assumption of socio-technical systems theories, namely that the organization of people and technology is interrelated and concern how functions are allocated between the two. His chapter, "Betwixt designs: streamlining or staying in the mess of practice?," aims at going beyond "supporting" work, since new technology changes work practices and their organizational setup. Such changes may actually be the goal of a new system, leading to new technologies "constituting" rather than supporting work and organizations. Hence, design and development of technology finds itself in a betwixt position concerning the purpose or purposes of the (re)distribution functions and responsibilities. Health care IT development, in particular, sets designers and developers of IT systems in betwixt positions—they have to choose whether to design for one or several purposes and may have to do stakeholder analysis to assess who has stakes in the envisioned IT system, as well as their ability to further or hinder its success.

Redesigning socio-technical systems can hence be seen as an intervention into practices with multiple purposes and stakeholders. The second sense in which designs are betwixt and which add an additional layer of complexity is to developing health care IT. This has to do with the way in which the perception and representation of work practices is by itself involves choice and can be seen as an intervention. Representations of practices should then not be made too rash and should build on detailed empirical knowledge. Two cases are analyzed with respect to health care IT development, concerning the development of a basic model for electronic health records (EHRs) in Denmark, and involving a logistic system for the coordination of hospital porter services.

Martina Augl's and Christian Stary's chapter addresses a major objective in the clinical operation of a hospital, namely, to ensure the availability of qualified personnel in daily operation. The quality of planning is particularly crucial when handling both stationary and walk-in patients. Daily scheduling is challenging due to the diversity of backgrounds and interests of the involved stakeholders. Various experts, such as doctors, nurses, technical support, and administrative staff, must find ways to collaborate for accurate planning. The approach taken in this study reveals that a dedicated, interaction-centered perspective on organizations allows for value networking before modeling and implementation of technological artifacts, in this case, subject-oriented business-process modeling and execution of validated models. Of crucial importance can be re-thinking the way stakeholders interact when accomplishing certain tasks, as it allows an organization to raise awareness of different planning conceptions. Focusing on the self-recognized interaction potential of involved stakeholders, existing work practice and thus IT support has been be significantly changed. The presented results demonstrate how interaction opportunities can be disclosed in a socially balanced way and prototypically implemented in a socio-technical setting for iterative refinement. Fundamental enablers were the actor-/communication-centered perspective on work processes and technology that allowed the 1:1 mapping of models embodying this perspective to interactive process experience.

Aleksandra Sarcevic, Ivan Marsic, and Randall S. Burd in their chapter "Challenges and Lessons Learnt on Dashboard Design for Improved Team Situation Awareness in Time-critical Medical Work" describe the design process of a clinical dashboard for improving team situation awareness during trauma resuscitation. It is a time-critical, high-risk, team-based, and information-intensive process of treating critically injured patients early after injury. The design approach was grounded in participatory design, allowing the authors to involve clinicians and domain experts in the system development, and to achieve common grounding across disciplines and among different stakeholders. The methodological approach included participatory design workshops, heuristic evaluation sessions, simulated resuscitations, video review of live resuscitations, and interviews. Stakeholders needed IT solutions meeting challenges of information access and retention, team coordination, and team situation awareness. In particular, synthesizing patient and process information was studied in this case, including ad-hoc and unpredictable work processes, with extremely dense timelines, high risk of human error, and diversity of information needs for interdisciplinary teams.

Xiaomu Zhou, Mark S. Ackerman, and Kai Zheng describe an ethnographic study at a large teaching hospital that examined doctors' use and documentation of patient care information, with a special focus on a patient's psychosocial information. The authors were particularly interested in the gaps between the work of the clinicians and the representations of the patient—specifically their psychological and social situations—in the EHR. The paper describes how doctors record information for immediate and long-term use. They found that doctors documented a considerable amount of psychosocial information in the EHR system; however, they also observed that such information was often recorded in a limited and too selective manner to be reused subsequently. The study shows how missing or problematic representations of a patient affect work activities and patient care not only in the present but also over time. The authors accordingly suggest that medical systems can be made more useful in the long run, by supporting representations of not only medical processes but also the patients themselves.

Michael Prilla and Thomas Herrmann report on lessons learned from designing collaborative reflection support in health care. They have worked in a German neurological hospital with a ward dealing with stroke patients. The ward was run by two senior physicians, who coordinated six to eight assistant physicians and the nurses of the ward. As a topic of research supporting physicians in learning about their conversations with relatives of patients had been chosen. Physicians had perceived a need to improve their skills in conducting these conversations, and that the lack of skills had created emotional stress and a bad reputation for the ward and the hospital. The project team has faced several challenges related to the health care domain that make the design of socio-technical support for reflection at work harder. These challenges included aspects of technology adoption, alignment to structures and processes in health care, technical support for solutions, and many more. The questions related to these challenges that they could contribute to are *what are the specific socio-technical design challenges related to healthcare*, and *how can stakeholders deal with them?* Some initial answers to these questions could be developed by comparing the work in the cases at hand to previous work on similar challenges, thus delivering outcomes that are applicable to a wide range of health care work places.

Jakob E. Bardram and Mad M. Frost in their chapter "Double-Loop Health Technology: Enabling Socio-technical Design of Personal Health Technology in Clinical Practice" examines chronic diseases that generally progress slowly requiring continuous care and treatment. Thereby, health care systems targeting personal and mobile health management can be of great support. Personal health technology is aiming at designing embedded sensor systems and using mobile and wearable computers for a novel pervasive, user-centered, and preventive health care model. The presented study of the MONARCA system for self-management and monitoring mood disorders reveals the overall design paradigm behind most of these applications establishing so-called "single-loop" treatment. Hence, single-loop applications are focused mainly on patient self management of wellbeing, healthy living, and/or disease care and treatment. "Double-loop" personal health technology, in contrast, involves both the patient as well as the clinician in the greater health care system. The authors explored this dual approach to upcoming personal health technologies. Of crucial importance is the opportunity to improve quality of the existing treatment as well as doing it in a more client-centric manner, aside introducing such technologies into the existing organization and practice of established health care systems in a professionally and socially acceptable way.

Overall, the work presented in the various cases shows that recognizing the social reality of system actors/users or even socio-ecological systems in health care is imperative, as it not only allows embodying this reality into processes but also to win over stakeholders and users as active participants in design. This book aims to provide insights into a variety of approaches to implement this perspective and should be used to inspire the design of further socio-technical solutions in healthcare. While each new case in health care will most likely be different from the ones presented here, practitioners, designers, and all other stakeholders involved may draw aspects and approaches from it and use them for their cases at hand.

Reference

Mumford, E., 1994. New treatments or old remedies: is business process reengineering really socio-technical design? The Journal of Strategic Information Systems 3 (4), 313–326.

Further Reading

Wulf, V., Rohde, M., Pipek, V., Stevens, G., 2011. Engaging with practices: design case studies as a research framework in CSCW. In: Proceedings of the ACM 2011 Conference on Computer Supported Cooperative Work, ACM, pp. 505–512.

Socio-technical Design for the Care of People With Spinal Cord Injuries

Mark S. Ackerman, Ayşe G. Büyüktür, Pei-Yao Hung, Michelle A. Meade, Mark W. Newman
University of Michigan, Ann Arbor, MI, United States

1. INTRODUCTION

Spinal cord injury (SCI) is a difficult, complex, and chronic condition. Injuries commonly result in paralysis and loss of normal function. Currently, there is no known cure. For those with an injury, managing one's health and mitigating secondary conditions is often physically and psychologically hard. Care must be maintained over one's lifetime.

Managing a SCI is complex and highly individualized (Hammond et al., 2009; Maddox, 2007). Each affected individual must master a range of self-care skills, including physical self-care, exercise, medication adherence, healthy eating, stress management, and emotional self-awareness (Meade and Cronin, 2012; Nunes et al., 2015). Mastering such a range of skills can be challenging, especially when patients leave the rehabilitation unit and have little access to professional support.

Much of the long-term burden of care falls on the patient and her family. Care can include help with continence and even breathing, help with the necessary exercises to maintain physical tone, and even making sure that helpers and supplies show up. Every patient is different and requires customized care at some level (Hammond et al., 2009).

Self-care is obviously centered on the patient herself. (Self-care and self-management tend to be used interchangeably in the Human-Computer Interaction literature; in this chapter, we follow Nunes et al. (2015) and use *self-care* to include so-called self-management tasks.) Assistance with care is often provided by a group of people we will call "caregivers" and will describe more fully below. This group includes spouses, parents, and siblings. Some families are able to hire aides, that is, health care helpers, people with relatively low-skill levels who can assist the individual and/or family in assisting with or completing required tasks at home. Helpers may come from an agency, but they may also be college students and volunteers. Family members are often prominent in assisting with care.

1

While SCI is a unique condition in some ways, its problems mirror other conditions. For example, the elderly often require similar care, although cognitive declines may limit their ability to direct care and participate in their self-management as fully as some patients with SCI. SCI is a particularly fruitful domain in which to develop and examine the potential for technical augmentation, due to the complex, collaborative nature of care and the strong need for customization to the individual.

To help people with SCI, we designed a system called SCILLS (pronounced "skills", short for the Spinal Cord Injury Living and Learning System), to be described below. From the beginning, it was designed as a technical system that had to fit in the specific social context of SCI. However, we did not anticipate how much that specific social context would interplay with the specific technical requirements. This chapter describes the rationale and design outcomes for SCILLS.

In the chapter, first we describe a standard scenario of use. We then follow with a description of our envisioned system, along with the design rationales for the system. We then describe our initial formative evaluation of the design. We learned a considerable amount from the formative evaluation, and after describing the evaluation, we survey the lessons learned as well.

2. SPINAL CORD INJURY

Individuals have different care needs based on the level of SCI and the completeness of injury. In SCI, the higher the level of injury (the closer the injury is to the neck area along the spinal cord) the more dysfunction the patient experiences and the more they require assistance with care activities. The spinal cord is enclosed by the spine (i.e., backbone), which is organized from top to bottom into the cervical, thoracic, lumbar, and sacral regions. Nerves originating at specific levels of the spinal cord go to specific areas of the body, each nerve exiting the spine between specific vertebrae. These nerves are numbered from top to bottom (one being at the top). Injuries are signified by their location: C1 through C8 (cervical or neck injuries), T1 through T12 (thoracic or upper-back injuries), L1–L5 (lumbar or mid-back injuries), and S1–S5 (sacral injuries).

For SCI, it is also important to note that injury may be "complete" or "incomplete." In a complete injury, there is no sensation or movement below the level of injury. In an incomplete injury, there is some function below the injury and the lack of function may not be symmetric; for example, there may be more movement in one limb than the other (What You Need to Know About Spinal Cord Injuries, n.d.). Both complete and incomplete injuries can occur at any level of the spinal cord.

Complete, cervical injuries cause quadriplegia (also called tetraplegia), including paralysis and loss of sensation in both the upper and lower extremities and the trunk. C1–C4 injuries (high cervical injuries) are the most severe. With the exception of some control in the neck, both the upper and lower body is affected. Individuals may not be able to breathe without a ventilator. People with high cervical injuries require 24-h assistance with care activities, although they may still be able to use powered wheelchairs on their own.

With C5–C8 complete injuries (low cervical injuries), there is more function in the upper extremities. In C5 injuries, there is shoulder and biceps control, but no control below the elbow. In C6 injuries, there is also wrist control but problems with the hands. At these levels people are able to do some self-care activities with assistance. With C7–C8 injuries, one becomes able to manage many self-care activities, but there may still be dexterity problems with the fingers. Individuals can get in and out of beds and wheelchairs without assistance. While relatively able to care for themselves independently (e.g., eat and dress), like with most spinal cord injuries they do not have bladder or bowel control. They need several hours per day of assistance with various self-care activities.

Thoracic (T1–T12) and lumbar (L1–L5) injuries cause paraplegia, including paralysis and loss of sensation in the lower extremities. People with paraplegia can use a manual wheelchair and modified car. They can be independent in self-care activities, including managing their bowel and bladder on their own, which at these levels of injury still lack normal function. People often need help with housework. Lumbar (L1–L5) injuries allow for some hip and leg control. People with these injuries can complete almost all tasks independently except for heavy housework.

We designed SCILLS to help SCI patients and caregivers.[1] SCILLS helps people with SCI acquire self-care skills through a clinician-managed virtual coaching program. Based on discussions with domain experts, we envisioned a particular use, which we next review.

3. SCENARIO

In a diving accident, Tim suffered a C6 SCI, meaning that he was unable to control his legs, hands, bladder, and bowels and had only limited control of his arms. He spent 2 months in the hospital, where he worked with rehabilitation clinicians.

As his discharge date approaches, Tim's nurse, Kevin, meets with Tim and his mother, Sharon, and introduces them to the SCILLS system. The nurse shows them how to create a Self-Care Plan (simply called a Plan in SCILLS) for Tim, which is a list of what he should do and how to carry out the activities. Kevin shows them how to track plan progress and access information digests. Kevin suggests some reminders, alerts, and triggers for bladder and bowel activities. These reminders, alerts, and triggers work on data being returned about what Tim is doing. Kevin also sets up information digests that would be useful for bladder and bowel management.

Upon returning home, Tim's mother, Sharon, helps Tim by setting up a plan using Kevin's recommendations. She adjusts the reminders about bladder and bowel management, knowing that Tim is a late riser. Sharon also selects two information digests that were recommended: one that provides several different stories about how individuals with SCI were able to find bladder management programs that worked for them, and one that discusses strategies for effective communication with one's health care provider. (In an actual scenario, there would be many more activities being done and facilitated.)

Throughout the weeks that follow, Sharon uses SCILLS to follow the plan and track progress and adds elements to the Plan, including items for performing skin checks, taking

[1] People with spinal cord diseases or some neuromuscular disorders have care needs that are closely comparable to those of individuals with SCI. Some individuals with conditions other than SCI were included in our work; however, we primarily report on those with SCI unless noted.

medication, and performing stretching exercises. She also fine-tunes the settings on her version of the self-care plan to optimize the effectiveness of the reminders and to lower the burden of data entry. After a month, it is time for Sharon to return to work, so she hires a crew of three home care helpers. They will work in shifts to help care for Tim. The helpers received basic training in home health care, but only one of them has experience specifically working with individuals with SCIs. Sharon trains the helpers, including how to use the SCILLs components, and nervously returns to work.

After a week, Sharon reviews the plan's progress and observes that medication and bowel management are on track but that, while Tim and the helpers were successful at performing bladder management according to schedule, they found wetness indicating bladder accidents. To understand this better, Sharon modifies the self-care plan to monitor Tim's fluid intake on an hourly basis while awake. When Sharon clicks the "accept" button, the modified self-care plan goes to Tim's care manager and physician for potential review.

While this scenario covers only a small portion of care, it illustrates key aspects of the envisioned use of SCILLS. Sharon could compose a customized self-care plan for Tim and his helpers based on her intimate knowledge of Tim's condition, personality, and lifestyle. Feedback about the aspects of the plan that worked well and those that did not helped Sharon tailor the plan in ways that could support Tim's progress. Other capabilities of SCILLS not highlighted in this scenario include providing alerts and data to clinicians, reuse of others' self-care plan elements, and the ability to author more complex plans including, for example, conditionals.

We next describe the SCILLS system.

4. SCILLS, THE SPINAL CORD INJURY LIVING AND LEARNING SYSTEM

In this section, we provide an overview of our initial design for SCILLS. SCILLS is currently in its second iteration of design, having gone through over a year of design, prototype construction, and formative evaluation.

Previous studies found that mobile software systems can help patients adhere to recommended treatment programs and develop self-care skills outside the clinical setting (Ding et al., 2010). However, the complexity of comprehensively managing a SCI or disease introduces new challenges. First, it is seldom feasible to try to master all needed skills immediately. Overwhelming the patient with tasks and responsibilities can increase stress and impede learning (Boschen et al., 2003; Fisher et al., 2002). This suggests that software to support skill development must be dynamic, adapting to the patient's level of mastery. Second, the complexity and individuality of different people's rehabilitation needs, environmental context, and care support system suggests that patients and clinicians will need to work together to customize and evolve self-care plans that work for the individual. For the patient to effectively participate in the design and implementation of their care program, it is critical that they understand not just *what* actions need to be taken, but also *why* and *how*.

4.1 Prior Work

A number of projects have explored using mobile technology to provide coaching and support to individuals with chronic conditions and disabilities. An application area that has received a great deal of attention is medication adherence (e.g., Choi et al., 2008; Lundell et al., 2007). While gains in adherence can be obtained through improved scheduling alone (Siek et al., 2010), it has been shown that leveraging user context, including location and activity, can improve the effectiveness of medication reminders significantly (Kaushik et al., 2008). Additionally, sensing technology can help with tracking medication adherence (McCall et al., 2010), playing a valuable role in evaluating the effectiveness of reminder systems as well as helping to moderate their operation. Systems to promote medication adherence, however, are just one component of the larger domain of *virtual coaching*, which comprises systems "aimed at guiding users through tasks for the purpose of prompting positive behavior or assisting with learning new skills" (Ding et al., 2010). While coaching systems have been explored most thoroughly in the domain of supporting people with cognitive impairments (e.g., Boger et al., 2006; Mihailidis et al., 2008; Pollack et al., 2003), they have also been examined in other domains related to physical rehabilitation (Liu et al., 2010) and management of chronic disease (Quinn et al., 2008). In their survey of virtual coaching systems, Ding et al. argue that the critical considerations when designing such systems include self-monitoring approach, context awareness, interface modality, and coaching strategy (Ding et al., 2010). Importantly, they argue that sensing and inference are critical enabling technologies for both assisted self-monitoring and context-aware reminders, both of which are important for the success of mobile virtual coaching systems.

The complexity and diversity of SCI manifestations argued that the personalization of virtual coaching protocols would be critical. While prior systems for reminding and coaching provided interfaces to allow users to initialize and update details of their prescriptions or care plans (McCall et al., 2010; Siek et al., 2010), these assume only simple data input would be required (e.g., dose and frequency). At another extreme, Autominder learns specific user activity patterns so that it can help the user accomplish them later when the user's cognitive function has been reduced (Pollack et al., 2003). This approach does not work in cases where the activities requiring assistance are new or where a potentially lengthy training period is not feasible. We favored a third approach, which seemed more promising for the domain of SCI coaching – end-user customization by the person with SCI and his care providers. The memory aiding prompting system (MAPS) adopts just such an approach, developing a "meta-design" environment in which caregivers with special knowledge of cared-for individuals with cognitive impairments are empowered to create custom "scripts" that allow the impaired person to perform personally meaningful activities of daily living (Carmien and Fischer, 2008). MAPS uses a "film-strip–based scripting metaphor" to allow care providers with little technical know-how to create complex multimedia scripts, though other metaphors have been explored to help nontechnical users create context-aware rules (Dey et al., 2006; Rodden et al., 2004; Truong et al., 2004).

Our review of the prior literature, preliminary interviews, and discussions with domain experts led us to the design of the SCILLS system, described next.

5. SYSTEM DESCRIPTION

The critical abstraction in the SCILLS system overall is a Care Plan. We describe that next, and follow with brief overviews of the important components of SCILLS—the care plan language, the Patient Coach (PCoach) application, and Clinician Builder (CBuilder) application (see Fig. 1.1).

At the heart of SCILLS are self-care plans (sometimes called care plans, self-management plans, or treatment plans). Self-care plans are sets of goals with a step-by-step breakdown of how to achieve those goals. Fig. 1.2 shows a sample, high-level self-care plan:

A physician may sketch out a self-care plan with relatively few details. These instructions are often elaborated by a lower-level clinician, such as an educator, nurse, or occupational therapist and might look like that in Fig. 1.2. As well, these lower-level clinicians may directly engage the patient with the necessities of their care. Regardless, each step in Fig. 1.2 would be further elaborated at some point—and perhaps multiple times as the patient and caregiver come to understand the condition and the clinicians come to understand the patient and caregiver.

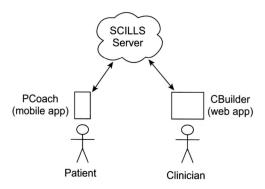

FIGURE 1.1   General architecture of the SCILLS system. *SCILLS*, spinal cord injury living and learning system.

7:30am – 8am	Wake up	
8am-8:30am	Cath (ISC); If in Bed, roll & reposition (R&R) If in chair, pressure relief	Catheter; gloves; urinal
8:30am-9am	Take medications; Eat breakfast If in chair, pressure relief	Medications
9am – 9:30am	Perform bowel program If in chair, pressure relief	Assistance to transfer to commode; gloves; suppositories
9:30am – 10am	Bowel program (continued)	
10am-10:30am	Get dressed and transfer to wheelchair ; Check skin	
10:30am-11am	If in chair, pressure relief	
11am-11:30am	If in chair, pressure relief	
11:30am-noon	If in chair, pressure relief	
noon-12:30pm	Cath (ISC): If in chair, pressure relief	Catheter; gloves; urinal
12:30pm – 1pm	Medications and lunch If in chair, pressure relief	
1pm – 1:30pm	If in chair, pressure relief	

FIGURE 1.2   High-level self-care plan for skin care, medication, and toileting (Meade, 2009, p. 34).

For SCILLS to provide sufficient assistance, using sensor-based monitoring and planned reminders, self-care plans have to be described in a way that can be automated. We began by designing a special purpose-language (Maloney et al., 2010), the Self-Care Plans for SCILLS (SCP4S) language. SCP4S is the glue for all of the components described above and provides the abstractions necessary for the performance of the overall system.

SCP4S includes characteristics of the three essential roles in the patient's situation: the patient herself, the caregivers, and clinicians. The language (a fragment of which can be seen in Fig. 1.3) had to be able to detail rules that would include:

- The activities that the patient would be learning or doing, including the necessary prompts and reminders for those activities,
- The measurement of those activities,
- The timing of those activities, and
- The feedback on the monitoring of those activities to the caregiver and/or a clinician

The language also had to include the ability for users to demark key elements of the patient's physical and sensor environment, since this is critical for monitoring. Finally, self-care plans also need to incorporate a limited amount of data about the patient's social environment (e.g., the roles and/or people who would be available as caregivers or clinicians). Note that we assumed that the self-care plans would be constantly evolving, and therefore the language had to be carefully constructed to make this easy.

Self-care plans, after being created and described in natural language and/or in the SCILLS plan language, have to be presented to the patient and/or caregiver. This is the role of the Patient Coach (PCoach) application. From the viewpoint of the user, PCoach is the most important part of SCILLS. It allows patients to view a schedule of activities, review information about their health conditions and recommended treatments, enter self-monitoring reports, and receive reminders for incomplete actions. Additionally, PCoach was designed to facilitate communication among the patient, caregivers, and clinicians, by allowing the patient or a caregiver to make changes to the schedule and plan (which are then communicated to the

FIGURE 1.3 This mockup shows a fragment of a self-care plan to prompt the user once a day to record his weight. The variable weight is defined elsewhere as being of type number, so the user will be prompted to enter a number. This request also recommends an information digest about healthy eating.

clinician for review), suggest information or videos about the activities, update plan progress by checking off completed actions, with optional comments, and enter questions, problems, or barriers encountered as a way to trigger further conversation with a care provider—perhaps at a later date. PCoach relies on a ContextEngine, which monitors contextual information provided by sensors.

In addition, SCILLS has an application, CBuilder, that allows clinicians to create and edit self-care plans, attach relevant information resources for patient education, monitor patient progress through self-monitoring reports and updates of completed actions, and respond to patient questions and problems.

6. FORMATIVE EVALUATION

Following the standard Human-Computer Interaction iterative cycle (Shneiderman et al., 2017), we wanted the SCILLS system to be designed, developed, and evaluated through interactions and repeated testing with both patients and clinicians, ensuring that the system would be developed and tested in the appropriate environment of use, making SCILLS be useful for and usable by members of the target audiences. The standard iterative design cycle used by many, if not most, HCI/CSCW projects has been found to be highly effective in creating usable and useful designs.

Therefore, we began with a two-fold approach, first studying some relatively simple tasks we wished to support: we chose (based on our readings on SCI care and initial conversations with patients, caregivers, and clinicians) hydration, exercise, and eating healthily. Drinking fluids is important for individuals with SCI, and we believed it could be measured using existing sensors such as smart water bottles. Drinking fluids is best monitored in conjunction with toileting, to make sure the two are in balance. Since we felt toileting would be harder technically, we examined hydration in detail. We also chose to consider pressure relief and pressure sores, as clinicians told us this was a critical activity that people did not always do. Pressure sores are caused by constant pressure on the skin from lying or sitting in one spot for a long time, and individuals with SCI do not check their skin as often as clinicians would prefer. At the same time as the formative evaluation, we developed a first-round prototype of the SCILLS language, SCP4S, as well as a simple plan editor.

What we found, in short, was that our initial assumptions, although aware of the importance of the patient and caregiver's social context along with their required coordination with clinicians, were naive and limited. In addition, as we began to construct our prototype language, we uncovered a number of limitations that were largely socio-technical in nature and which deepened our understanding of the requirements for the SCILLS system.

Below we discuss the findings from the interviews. We will then follow that by unpacking the contextual issues for the technical system.

6.1 Examination of Activities

SCI causes chronic health problems that must be monitored and managed on a daily basis. We conducted 21 interviews with people with SCI, caregivers, and clinicians (plus another three with people dealing with related conditions). Our interviews attempted to determine the basic issues in the activities, any problems routinely encountered, and the feasibility of helping patients or caregivers with those activities. As mentioned, we focused on three common activities, but the interviews ranged over other activities as well. The interviews were approximately 1 h in length, and with permission, were audio-recorded and transcribed. We used Clarke's Situational Analysis (Clarke, 2005), an updated version of Grounded Theory, to analyze the data. Any participant identities or quotes used here have been made anonymous.

6.1.1 Bladder and Bowel Functions

People with SCI are usually unable to control bladder and bowel functions at will. They must develop programs to empty the bladder and bowel regularly to avoid unwanted accidents and potentially serious complications. Effective bladder and bowel programs are intrinsically related to careful management of hydration and nutrition as well.

The fear with both bowel and bladder programs is that an accident will occur at an unwanted time or place unless the activities are regularly scheduled and successfully completed. Larry, who has been a peer mentor to other individuals with SCI for many years, noted that one of the topics newly injured people are most interested in is the bowel program. This is due to psychosocial reasons, and because it involves trying different approaches until a favorable routine is established:

> ...the bowel program I think is the hardest because sometimes it doesn't work. And no matter what you do it's not working. Whether you're using a suppository, or you're doing digital stimulation, or you're doing both, and you [have to] invest more than an hour or two into just going to the bathroom.

As Larry mentioned above, in the case of a bowel program there is the need to establish a process for bowel stimulation (via medications or other means) that is time consuming. A routine can take anywhere from 30 to 60 min to several hours to complete. In addition, for those with higher level injuries, a bowel program usually necessitates the help of another person and must therefore be scheduled at a time when a caregiver is available. Family members can and do learn how to help with bowel programs, but some people prefer help from a hired caregiver because of the highly personal nature of the activity. For example, Tina—a caregiver—stated that her son with quadriplegia (tetraplegia) specifically requested that his parents not get involved, delegating the task to a hired caregiver. On the other hand, Tina also noted that even with caregivers the bowel program "is a *big* issue. Because that's something obviously nobody has been used to doing."

Megan, an individual with quadriplegia—explained how her bowel routine developed through trial and error once she got home from the hospital after rehabilitation. She eventually arrived at a routine that has been working for her very effectively:

> [I]n my situation I have to be a little more resourceful.... You take a little juice basically, you take it in a little tube, squirt it in there and then in about 30 min you're going. And it takes care of it *so* much easier than anything else.

Much like the bowel program, many people tend to routinize the bladder program early on as well. Bladder-care routines also evolve over time through intermediate, increasingly favorable routines that bring stability to the activity. Commonly, catheterization is at first done by a caregiver, and people try to work out the optimal schedule to avoid accidents. Megan explained that in the beginning her mother usually did her catheterization. Later on, Megan decided she could learn how to catheterize herself despite having dexterity issues with her hands:

> And then the other thing was cathing. Because at the time when I got home I wasn't cathing myself… And early on [my mother] was … the one doing a lot of it, which was, you know, get me up in the morning, cath me. …But as time went on I learned I can do it and did it.

6.1.2 Hydration

Our participants noted that proper hydration is crucial for a number of reasons, including avoiding fatigue or complications such as urinary tract infections. People with SCI are generally aware of the importance of hydration and develop routines to try to maintain sufficient fluid intake.

Recommendations by experts concerning the amount of daily hydration vary considerably. Some professional societies recommend 2–3 L a day, an SCI nurse on a forum commented that the human body needs about 9 cups of fluid for women and 13 cups of fluid for men, and Greg—a Physical Medicine and Rehabilitation (PM&R) doctor we interviewed—explained how the amount of hydration has to be personalized:

> Most of that stuff that's in the general public knowledge about hydration is just inaccurate… You know, it would be really dependent on their bladder program… What are their volumes? What's the frequency of their catheterization? If they're quadriplegic, are they having any symptoms of dysreflexia? And then are they having any issues related to urinary tract infections? Then also like I said, accidents.

In the quote above, the doctor points out some of the complexities in hydration routines (bladder output and frequency of catheterization), as well as some complications that commonly disrupt these routines (e.g., autonomic dysreflexia, urinary tract infections). Hydration, in general, is quite contextualized. It depends on the types of fluids a person drinks (e.g., alcohol, juice, or water), the ambient temperature (hot or cold), and the amount of activity the person is engaged in. Instead of measuring, people with SCI often inspect the color of their urine to determine whether they are properly hydrating.

We found that people develop hydration routines over time that are stable, but that these routines have to be reconstructed or recreated if there is a disruption. For instance, a common response to a urinary tract infection is to increase hydration to help clear the infection. Once the infection clears, people may revert back to their original routine. They may also permanently change their routine, although whether or not the new routine is more effective than the original may not be readily apparent. For instance, evidence from SCI forums indicates that for some people drinking cranberry juice or taking cranberry pills helps to prevent urinary tract infections, while for others these turn out to be ineffective. On the other hand, at times routines change more permanently after a disruption. Coronary problems tend to require individuals to decrease their fluid intake, as do bladder accidents that may happen.

6.1.3 *Pressure Sores*

As mentioned, pressure sores are caused by constant pressure on the skin from lying or sitting in one spot for a long time. Individuals with SCI are particularly disposed to getting these as they are often in bed or in a wheelchair. A pressure sore may vary from a red spot on the skin to a deep wound down to the bone. If a sore does not heal properly it can lead to severe complications (e.g., requiring amputation) and even death.

Pressure relief involves the shifting of one's weight on a regular basis, which is critical to prevent the formation of pressure sores. The standard recommendation is to do pressure relief at least every 15 min during daytime. Greg, a PM&R doctor, stated that uninjured individuals unconsciously "squirm" to shift their position frequently. Individuals with SCI can do pressure relief themselves by tilting backward and forward in the wheelchair, but it is not an unconscious activity for them because of the nerve damage caused by injury. They often do not feel discomfort or pain from constant pressure, and they also cannot see for themselves that a pressure sore is beginning to emerge because these are usually located on the back of the body. They must therefore remember to complete this activity regularly, throughout the day.

We found that pressure relief is one of the most neglected activities despite clinicians' warnings. In addition to clinicians' observations from experience, individuals with SCI and caregivers noted that it is common to go several hours without doing any pressure relief. James, the father of an individual with quadriplegia, noted that his son would sit for hours without moving, only to realize he has not done pressure relief when something causes him to "spasm out." Other caregivers and individuals with SCI, as well as clinicians, have noted that people often do not do pressure relief as instructed. However, disruptive reminders provided every couple of minutes are not deemed as an effective mechanism to help routinize this activity. Megan, for instance, commented that she would be "annoyed" if she were constantly interrupted, even by a caregiver. It is indeed the case that many individuals default to completing this activity on an ad hoc basis instead of setting alarms to remind them every few minutes.

6.1.4 *Other Activities*

There are many other issues for people with SCI including problems with breathing and other respiratory issues (sometimes requiring the use of ventilators and cough assistance), as well as muscle stiffness and involuntary movements (spasticity). Sleep hygiene and chronic pain are often problems, and depression and isolation can be issues. Over time, the sedentary lifestyle, as with other people, may lead to bone density loss, obesity, cardiovascular disease, and diabetes. The muscle problems may lead to a limited range of motion. As well, for individuals with paraplegia, shoulder injuries mount over time, since the shoulders are strained with manual wheelchair use.

In addition, each person with SCI must manage a number of activities, either independently or with the help of caregivers, related to physical care (e.g., skin care, bladder, and bowel management), medication adherence, exercise, nutrition, and hydration. Besides the common issues described above, everyday activities such as feeding, bathing, grooming, and household tasks may require considerable effort and potentially caregiver help. Transfers (e.g., between a bed and a wheelchair) may be required depending on the person's needs.

6.1.5 *The Social Context of Care*

These activities are not usually done in isolation. People with SCI work with family or hired caregivers, critically influencing the management of self-care activities. The age of the injured person and time after a traumatic injury affect dependence on caregivers: children and those who are newly injured tend to rely more heavily on caregivers. Individuals gradually take more responsibility for self-care, although reliance on caregivers for certain activities may be permanent depending on the level of injury. Caregiver arrangements vary and are often associated with the availability of funds. Family members are usually closely involved in the care process regardless, with parents and spouses taking the lead. However, hiring outside caregivers for a few hours per day or per week is common.

In addition, clinicians continue to oversee medical care and provide recommendations—which become part of the self-care plan—on an outpatient basis. Notably, for those with access to specialty centers, doctors specializing in physical medicine and rehabilitation, urologists, occupational therapists, physical therapists, rehabilitation psychologists and rehabilitation engineers continue to be centrally involved. Depending on the needs of the individual, other specialists (e.g., respiratory therapist, dietitian, social worker) may also be involved.

At the same time that their care team is a network of people, a key concern of many individuals with SCI is to manage as much of their care as possible to become as independent as possible. However, mastering the knowledge and skills to achieve maximum independence can be challenging, especially when individuals leave the hospital after months of rehabilitation that immediately follow their injuries. Once they leave the hospital, individuals with SCI and caregivers tend to develop care routines that they prefer and stay with them.

In summary, activities done by people with SCI about their own care are often highly situated. To a large extent, this is the norm—people develop stable routines that work well enough and can readily adapt to the contingencies at hand. These routines therefore deal with the basic situational context of activities. For the three activities examined, the routines may need to be adjusted (and often are), but they are not constantly changed except under unusual circumstances. (Sleep and pain may cause substantial changes to routines to find new and even temporary solutions.)

7. TECHNICAL LESSONS

As we conducted our formative interviews, we also constructed several prototypes of our language and editing system. This language, as mentioned, was to be the glue between the coaching and clinicians. It tied together the kinds of data desired, the desired functioning of sensors, and the actions to be taken with sensor and user data.

We found that we could construct such a language; however, we believe that one would need to be careful in its uses. In short, if there is a gap between the capabilities of available sensors and the complex, situated social context, then the software that maps sensed values to system actions also explodes in complexity. This is especially true when the software must make inferences based on data that is incomplete or partial. Furthermore, any system employing noisy sensors in the tracking of users with clinical oversight (as was originally envisioned for SCILLS) runs the risk of producing more errors than patients, caregivers, or clinicians would tolerate.

While constructing a technical infrastructure to support the social requirements of use, including measurement, calculation, and alerting, we found four rough categories based on the adequacy of sensors for the activity in question:

- Activities for which there exists a sensor that accurately measures that activity. Completely accurate sensors that measure the activities of importance are currently rare. Even many simple activities either do not have the right sensors available for their detection and measurement, or the activities are sufficiently simple that people can do them largely without the need for clinical intervention. However, our participants were all highly motivated, and it may be that automatic monitoring could help those who are less motivated. Furthermore, automatic monitoring may help look for breakdowns people are having in an activity (e.g., problems with sleep).

  An example of a sensor that (very nearly) solves the hydration problem is Uchek.[2] This sensor can detect urine color, which is the basis for the relatively straight-forward heuristic that people with SCI use. Uchek simply measures the color of one's urine based on taking a urine sample, testing it with a known color strip, and then holding a cellphone up to the color strip. As noted above, the standard heuristic people with SCI are trained to use and do use is to note the color of the urine. If it is sufficiently dark, the person with SCI is not hydrating adequately. Although we have not yet tested this, the heuristic should be relatively straight-forward to program, given a sample, color strip, and cellphone.

  Note that the use of this sensor would require the participant or caregiver be motivated enough to collect the data in this fashion reliably and over time. This is particularly an issue since it is simpler to not avoid getting a urine sample and using a color strip. Nonetheless, one could easily imagine this sensor leading to a system like our goal—relatively easy to use and a highly reliable instrument.

  In this case, our language and system merely needs to provide clinicians with a template to set alert conditions. For people with SCI and their caregivers, they may wish a visualization application to see how they are doing.

- There is a class of activities for which no sensor can exist by definition, because at least part of the activity or state is subjective. Example conditions include pain and sleep. While sleep duration can be adequately measured with sensors, subjective sleep quality cannot. Pain is highly subjective. In these cases, an augmentative system might be able to partially monitor the person's ongoing conditions, but some amount of manual data entry will be required. This condition has the standard motivational issues of on-going data entry found in personal informatics applications (Epstein et al., 2015; Li et al., 2010). People with SCI and their caregivers may wish a visualization application to see how they are doing. Clinicians may find it more straight-forward to directly ask about the condition during medical appointments, although for people who see their physician only yearly, distance monitoring, and/or alerting might still be valuable.

[2]https://www.wired.com/2013/02/smartphone-becomes-smart-lab/.

- There is a class of situations where no sensor exists to reliably monitor an activity, but manual data entry can be simple. In some situations, just asking whether the activity has occurred may be adequate. An example of this is checking for pressure sores. Manual data entry could occur from caregivers or from the patient, and if reliably done, this would be adequate for distance monitoring. (Of course, in this situation the question of reliability is paramount.)
- No sensor can adequately measure an activity, even though the condition is not subjective. An example of this might be hydration before the Uchek sensor. This is an interesting condition to explicate, since it shows the complexity of using a combination of sensors to measure an activity.

Initially we attempted to use the MyHydrate water bottle[3] as our sensor for hydration. The MyHydrate water bottle promised to be "smart," but its functionality at the time was to simply measure how many fluid ounces it contained. Using any water bottle like it to measure hydration would place severe restrictions on the person with SCI. To have adequate measurement, he/she would need to take all fluids through the smart water bottle. Alternatively, we would have to find ways to wire each cup and other drinkware, which is not currently feasible. Thus this approach failed to achieve *sensor completeness*.

Even if this smart water bottle or other smart drinking utensils were used and the amount of fluid could be measured adequately, a considerable amount of data entry would still need to be done by hand. One might want to drink outside of one's standard utensils; the person with SCI might drink a bottled soda or beer while out with friends. As well, different drinks have different properties. The substance might be a diuretic, for example, tea. Diuretics have very different hydration properties from water (they dehydrate), so that information would need to be entered by hand or any calculations of hydration would be incorrect. Thus this approach also failed to achieve *data completeness*.

In addition, the calculation of hydration from the input side is far more complex than measuring the output. Even if data collection proceeded as one hoped, confounding issues might include whether the person left the house and went to a store by car, encountering four additional temperature exposures of various durations (home to car, car, car to store, store, with repetitions). In the depth of a cold winter or hot summer, these changes can be significant. Moreover, the calculations would need to be personalized to the person's bodily characteristics. Every person's metabolism is different. This approach also failed to achieve *computation completeness*.

One might contrast the complexity of automatic activity inferencing and calculation with how relatively easy it is for people with SCI to calculate hydration level as part of their normal practice. Much care goes into teaching the patient what this means in specific situations and what to do about it. Like Goodwin's (1994) competent practitioner, a patient (and any caregivers) must be taught to work out hydration and dehydration. They are taught, or come to understand, the color of the urine, and how to contextualize that including how to consider the ambient atmospheric temperature and other aspects of diet (e.g., an alcoholic drink while socializing). An alert, for example, to drink 200 mL every 2 h would be robotic and not situated. Even with the best of intensions, such an alert is likely to be ignored, since visual inspection would inform the patient that hydration is a problem or not.

---

[3] https://www.myhydrate.com/.

To summarize, we have not yet found a situation where meaningful self-care activities can be measured automatically with a strong level of reliability, a high level of usability, and/ or a level of clinical comfort. We have found situations, however, where a combination of manual and sensor-based monitoring could lead to a better understanding of people's activities. Because this is a research project, we have bracketed off question of installation and maintenance (e.g., ongoing calibration, network installation) from concern.

8. REFLECTIONS ON SOCIO-TECHNICAL DESIGN

Our project demonstrates important characteristics about the beginning of socio-technical efforts. Designing a socio-technical system is to design to a moving target. Partly, it is an effort to understand the social environment as a set of ongoing negotiated and constructed practices. For us, our design had to take into account the care practices and the kinds of caregiver networks and so we focused our efforts there. When designing, initially the social environment is relatively stable and appears as a set of constraints on the design. But as well, designs have to also understand the technical possibilities, either of particular systems or of technical environments. The technical systems are often givens, but for us, they could also be constructed. (The sensors lay outside of our design capabilities and therefore are given rather than under our control.)

Therefore, the story of SCILLS to date has been one of trying to understand the current practices of SCI patients, caregivers, and clinicians, as well as the capabilities and constraints of potential sensors and software platforms. As one can easily deduce, the project (at the time of writing) is at an early stage of development.

The state of understanding for SCILLS currently—where the details of the technical platforms and potential capabilities are still becoming clear while the complexity of the social context has been detailed—is not uncommon in socio-technical design. The considerations obviously differ from project to project, but uncovering and understanding the co-design space is common. The hard problem in socio-technical design is not in understanding the requirements of multiple, conflicting, or overlapping social contexts or the technical capabilities of various system components—although these are very difficult—but understanding *how the two will join together over time*. (See the analysis in Ackerman, 2000 for why this is difficult.). That is, one must understand the *socio-technical trajectory problem* in design. If one assumes there is an embeddedness (Bjørn and Østerlund, 2015) or entanglement (Barad, 2007) between the social context and artifacts (where the social context includes but is not limited to specific practices), then clearly there is a state where some artifact has not yet been entangled in a specific group's practices and then another state where it has been. How one gets from a separate social and technical to this embeddedness or entanglement needs to be studied.

Developing theories of socio-technical design cannot be limited to understanding how things became the way they are. In retrospect, the decision points in designs seem obvious or at least understandable. Guiding design is important. Few studies examine the messiness of design in the early stages. At the beginning of design projects, there are technical capabilities and there are social practices, but they have not yet combined. The difficulty in the design process is finding the new possibilities within the constraints that either exist or could exist.

While several academic areas look at socio-technical design, none have a good answer about how to carry out this step. For example, the science, technology, and society academic area looks at how systems relate to their social contexts, often at a macro-scale (but see e.g., Jackson et al., 2012; Vertesi, 2008). The classic Latour study Aramis (Latour and Porter, 1996) describes the socio-technical trajectory problem in the context of the history of a high-speed train design. HCI and CSCW look at what people do and believe strongly in iterative cycles of analysis and design. HCI and CSCW have an assumption that the social and the technical come to be intertwined, but they do not study the process by which that intertwining occurs. A newer area, found intertwined within CSCW and in organizational behavior and communications, calls itself socio-material studies (Leonardi and Barley, 2008) and is concerned with what we call socio-technical design here. For socio-material studies, technical capabilities enable or constrain changing social practices, and changing social practices drive new technical investigations and design. This newer area aims to understand the design process better. The theoretical framework is appropriate, but to date only some studies have studied how to formulate the early stages of design. Bjørn and Østerlund (2015), for example, examine designing medical practices, argue that one should explicate the bindings between artifact and practices and then systematically relax and tighten the bindings.

We cannot offer any complete solutions here; indeed, we found ourselves casting about for a method to rationalize our design process. However, in retrospect, creating a detailed matrix of technical affordances offered by the sensors (i.e., what they did) along with the social requirements should have been our first step. In our case, there were social requirements that were absolutely required and some that were preferred. In many cases, from a usability perspective, the care practices of the people with SCI and their caregivers cannot be easily changed because their lives are very dependent on the continuity of their routines. It might be possible to substitute care practices, but this would be a substantial effort. On the other hand, the care practices of clinicians, that is the kinds of alerts and data they might receive, are strongly preferred and in most cases required. For both, it was possible to try new practices within a prototype so as to examine their future potential. When we finally sat back and created such a matrix, we discovered that we could not adequately support the care practices with the sensors that were available, but with some additional manual data entry, we could provide alerting and tracking facilities. We are now constructing such a facility.

In conclusion, SCILLS was initially based on our preliminary analysis of the needs of people with SCI as well as our understanding of the current state of the technical art, as might have been expected. As we carried out the project, as described in this chapter, we discovered how far apart the social and the technical still were. This is also as expected. The details matter. What we want to highlight here is the process of moving from a general conception of the social requirements and technical affordances to having experience with the details. HCI/ CSCW has methods for pulling out the details of the social context; we followed them here. It still needs methods for pulling out the details of the technical environment.

Acknowledgments

The authors thank the TIKTOC Advisory Council and the SocialWorlds and Interaction Ecologies research groups for their support. The authors are grateful to all their participants for sharing their experiences. This project was funded by the Craig H. Neilsen Foundation (grant #324655) and the National Institute on Disability, Independent Living, and Rehabilitation Research (grant #90RE5012).

References

Ackerman, M.S., 2000. The intellectual challenge of CSCW: the gap between social requirements and technical feasibility. Human-Computer Interaction 15 (2–3), 179–204.

Barad, K., 2007. Meeting the Universe Halfway: Quantum Physics and the Entanglement of Matter and Meaning. Duke University Press, Durham, NC.

Bjørn, P., Østerlund, C., 2015. Sociomaterial-Design: Bounding Technologies in Practice. Springer International Publishing, Cham, Switzerland.

Boger, J., Hoey, J., Poupart, P., Boutilier, C., Fernie, G., Mihailidis, A., April 2006. A planning system based on Markov decision processes to guide people with dementia through activities of daily living. IEEE Transactions on Information Technology in Biomedicine 10 (2), 323–333.

Boschen, K.A., Tonack, M., Gargaro, J., September 2003. Long-term adjustment and community reintegration following spinal cord injury. International Journal of Rehabilitation Research 26 (3), 157–164.

Carmien, S.P., Fischer, G., 2008. Design, adoption, and assessment of a socio-technical environment supporting independence for persons with cognitive disabilities. In: CHI '08. Proceedings of the SIGCHI Conference on Human Factors in Computing Systems, Florence, Italy. ACM, New York, pp. 597–606.

Choi, J.-H., Lim, M.-E., Kim, D.-H., Park, S.-J., 2008. Proactive medication assistances based on spatiotemporal context awareness of aged persons. In: EMBS '08. Proceedings of the 2008 Annual International Conference of the IEEE Engineering in Medicine and Biology Society, Vancouver, 20 August–24 August, 2008. IEEE, Piscataway, NJ, pp. 5121–5124.

Clarke, A., 2005. Situational Analysis: Grounded Theory After the Postmodern Turn. SAGE Publications, Thousand Oaks, CA.

Dey, A.K., Sohn, T., Streng, S., Kodama, J., 2006. iCAP: interactive prototyping of context-aware applications. In: Fishkin, K.P., Schiele, B., Nixon, P., Quigley, A. (Eds.), PERVASIVE'06. Proceedings of the 4th International Conference on Pervasive Computing, Dublin, Ireland, 07 May–10 May, 2006. Springer-Verlag, Berlin, pp. 254–271.

Ding, D., Liu, H.-Y., Cooper, R., Cooper, R.A., Smailagic, A., Siewiorek, D., February 2010. Virtual coach technology for supporting self-care. Physical Medicine and Rehabilitation Clinics of North America 21 (1), 179–194.

Epstein, D.A., Ping, A., Fogarty, J., Munson, S.A., 2015. A lived informatics model of personal informatics. In: UbiComp '15. Proceedings of the 2015 ACM International Joint Conference on Pervasive and Ubiquitous Computing, Osaka, 07 September–11 September, 2015. ACM, New York, pp. 731–742.

Fisher, T.L., Laud, P.W., Byfield, M.G., Brown, T.T., Hayat, M.J., Fiedler, I.G., August 2002. Sexual health after spinal cord injury: a longitudinal study. Archives of Physical Medicine and Rehabilitation 83 (8), 1043–1051.

Goodwin, C., September 1994. Professional vision. American Anthropologist 96 (3), 606–633.

Hammond, Margaret, C., Stephen, C., 2009. Yes, You Can!: a Guide to Self-care for Persons with Spinal Cord Injury. Paralyzed Veterans of America, Washington, DC.

Jackson, S.J., Pompe, A., Krieshok, G., 2012. Repair worlds: maintenance, repair, and ICT for development in rural Namibia. In: CSCW '12. Proceedings of the ACM 2012 Conference on Computer Supported Cooperative Work, Seattle, Washington, 11 February–15 February, 2012. ACM, New York, pp. 107–116.

Kaushik, P., Intille, S.S., Larson, K., 2008. Observations from a case study on user adaptive reminders for medication adherence. In: PervasiveHealth '08. Proceedings of the 2nd International Conference on Pervasive Computing Technologies for Healthcare, Tampere, Finland, 30 January–1 Febuary, 2008. IEEE, Piscataway, NJ, pp. 250–253.

Latour, B., Porter, C., 1996. Aramis, or, the Love of Technology. Harvard University Press, Cambridge, MA.

Leonardi, P.M., Barley, S.R., April 2008. Materiality and change: challenges to building better theory about technology and organizing. Information and Organization 18 (3), 159–176.

Li, I., Dey, A., Forlizzi, J., 2010. A stage-based model of personal informatics systems. In: CHI '10. Proceedings of the SIGCHI Conference on Human Factors in Computing Systems, Atlanta, Georgia, 10 April–15 April, 2010. ACM, New York, pp. 557–566.

Liu, H.-Y., Cooper, R., Cooper, R., Smailagic, A., Siewiorek, D., Ding, D., Chuang, Fu-C., January 2010. Seating virtual coach: a smart reminder for power seat function usage. Technology and Disability 22 (1), 53–60.

Lundell, J., Hayes, T.L., Vurgun, S., Ozertem, U., Kimel, J., Kaye, J., et al., 2007. Continuous activity monitoring and intelligent contextual prompting to improve medication adherence. In: EMBS '07. Proceedings of the 2007 Annual International Conference of the IEEE Engineering in Medicine and Biology Society, Lyon, France, 23 August–26 August, 2007. IEEE, Piscataway, NJ, pp. 6286–6289.

Maddox, S., 2007. Paralysis Resource Guide. Christopher Reeve Foundation, Short Hills, NJ.

Maloney, J., Resnick, M., Rusk, N., Silverman, B., Eastmond, E., November 2010. The scratch programming language and environment. ACM Transactions on Computing Education 10 (4), 16 1–16:15.

McCall, C., Branden, M., Zou, C.C., Zhang, J.J., 2010. RMAIS: RFID-based medication adherence intelligence system. In: EMBS '10. Proceedings of the 2010 Annual International Conference of the IEEE Engineering in Medicine and Biology Society, Buenos Aires, Argentina, 1 September–4 September, 2010. IEEE, Piscataway, NJ, pp. 3768–3771.

Meade, M.A., 2009. Health Mechanics: Tools for the Self-management of Spinal Cord Injury and Disease. University of Michigan, Ann Arbor, MI.

Meade, M.A., Cronin, L.A., 2012. The expert patient and the self-management of chronic conditions and disabilities. In: Kennedy, P. (Ed.), The Oxford Handbook of Rehabilitation Psychology. Oxford University Press, New York.

Mihailidis, A., Boger, J.N., Craig, T., Hoey, J., November 2008. The COACH prompting system to assist older adults with dementia through handwashing: an efficacy study. BMC Geriatrics 8 (1), 28.

Nunes, F., Verdezoto, N., Fitzpatrick, G., Kyng, M., Grönvall, E., Storni, C., December 2015. Self-care technologies in HCI: trends, tensions, and opportunities. ACM Transactions on Computer-human Interaction 22 (6), 33 1–33:45.

Pollack, M.E., Brown, L., Colbry, D., McCarthy, C.E., Cheryl Orosz, Bart Peintner, et al., September 2003. Autominder: an intelligent cognitive orthotic system for people with memory impairment. Robotics and Autonomous Systems 44 (3–4), 273–282.

Quinn, C.C., Sysko Clough, S., Minor, J.M., Lender, D., Okafor, M.C., Gruber-Baldini, A., June 2008. WellDoc TM mobile diabetes management randomized controlled trial: change in clinical and behavioral outcomes and patient and physician satisfaction. Diabetes Technology & Therapeutics 10 (3), 160–168.

Rodden, T., Crabtree, A., Hemmings, T., Koleva, B., Humble, Jan, Karl-Petter Åkesson, Hansson, P., 2004. Between the dazzle of a new building and its eventual corpse: assembling the ubiquitous home. In: DIS '04. Proceedings of the 5th Conference on Designing Interactive System, Cambridge, MA, USA, 01 August–04 August, 2004. ACM, New York, pp. 71–80.

Shneiderman, B., Plaisant, C., Cohen, M., Jacobs, S., 2017. Designing the User Interface: Strategies for Effective Human-Computer Interaction. Pearson Education, Upper Saddle River, NJ.

Siek, K.A., Ross, S.E., Khan, D.U., Haverhals, L.M., Cali, S.R., Meyers, J., October 2010. Colorado care tablet: the design of an interoperable personal health application to help older adults with multimorbidity manage their medications. Journal of Biomedical Informatics 43 (5 Suppl.), S22–S26.

Truong, K.N., Elaine, M.H., Abowd, G.D., 2004. CAMP: a magnetic poetry interface for end-user programming of capture applications for the home. In: Davies, N., Mynatt, E.D., Siio, I. (Eds.), UbiComp '04. Proceedings of 6th International Conference on Ubiquitous Computing, Nottingham, UK, 7 September–10 September, 2004. Springer, Heidelberg, pp. 143–160.

Vertesi, J., February 2008. Mind the gap: the London underground map and users' representations of urban space. Social Studies of Science 38 (1), 7–33.

What You Need to Know About Spinal Cord Injuries. (n.d.). Johns Hopkins Medicine. http://www.hopkinsmedicine.org/healthlibrary/conditions/physical_medicine_and_rehabilitation/spinal_cord_injury_85,P01180/. Accessed 25 March 2017.

2

Design Principles for Supporting Patient-Centered Journeys

Maia Jacobs, Elizabeth D. Mynatt
Georgia Institute of Technology, Atlanta, GA, United States

1. INTRODUCTION: HEALTH CARE AS A JOURNEY

A central goal of ubiquitous computing has been to support individuals' everyday activities. As discussed in Abowd and Mynatt's seminal work (Abowd and Mynatt, 2000), people's everyday activities have no definitive beginning or end. Rather, activities such as information management and communicating with others are continuous and constantly evolving. Research in the area of everyday computing calls for designers of ubiquitous technologies to consider how we may scale systems with respect to time.

In the context of health, the goal of supporting these continuous activities is ever present due to the rise of chronic illness diagnoses worldwide. Today, chronic illnesses affect hundreds of millions of people. For those living with a chronic illness, daily activities such as exercise, monitoring side effects, and communicating with others can be critical to one's physical health and overall quality of life and can directly influence clinical outcomes. In response, international organizations are calling for an increased focus on supporting these self-management activities to improve the prevention and management of chronic illnesses (World Health Organization, 2005; Buntin et al., 2010).

In this chapter, we argue that time is a critical dimension to consider in socio-technical research and the design of personal health technology. We utilize the metaphor of a journey to represent both activities and time as factors within a health care experience. Similar to the definition of activity used within the ubiquitous computing community, health-related activities are made up of people and tasks, as well as the context in which they exist, with no conclusive beginning or end point. Therefore a critical challenge for personal health technologies is to provide the continuity and capacity necessary to support one's evolving health care journey.

We utilize the breast cancer journey, most prevalent cancer among women worldwide, as a case study for investigating how technology may better support these health care experiences. Existing work has described the cancer journey as it relates to treatment, highlighting common phases such as screening and diagnosis, acute treatments, and no evidence of disease (Hayes et al., 2008). However, to more broadly support an individual's cancer experience, we employ a socio-technical perspective to extend the notion of a cancer journey, using the patient experience as the focal point. We use this viewpoint to develop tools that consider individuals' range of health management tasks and challenges patients face throughout treatment and survivorship.

Over a 4-year period, we partnered with the Harbin Cancer Clinic, Rome Cancer Navigators, and the Northwest Georgia Regional Cancer Coalition in Rome, Georgia to study existing care practices and to develop technologies to help breast cancer patients throughout their cancer journeys. We began by studying the Rome Cancer Navigators' work practices to understand human processes for supporting cancer patients throughout treatment and survivorship. By understanding how people within the health care setting support patients' changing needs, we are able to identify effective methods used to provide support, as well as opportunities for technological implementations to complement these existing processes. We also worked with patients to better understand their experiences and how the cancer journey affects their daily lives and how individuals' needs and priorities changed over time. Using the results of these formative studies, we developed a tablet-based tool designed to offer flexible, comprehensive support to breast cancer patients as they progress through treatment and into survivorship. We evaluated this work by assessing the influence of the technology on patients' experiences and also on the indirect effects of cancer navigation practices. In this chapter we discuss the results of this work, which includes the following contributions:

1. We identify strengths in the existing cancer care socio-technical system, with a specific focus on cancer navigation practices and discuss opportunities to learn from and enhance these work practices through technological support, while addressing existing system challenges.
2. We ground our understanding of the cancer care system in the patient's experience through our work with cancer survivors, providing a robust qualitative assessment examining how patients' cancer journeys change over time.
3. We describe findings from a pilot technology study, known as the My Journey Compass project, which allowed us to discover how flexible tools can support patients' needs throughout treatment and survivorship.

We conclude this chapter with a reflection of lessons learned to guide the development of systems that support longitudinal and dynamic health care experiences. We developed these guidelines to extend beyond breast cancer, focusing on how systems may better support dynamic health experiences more generally.

2. BACKGROUND: PERSONAL HEALTH MANAGEMENT CHALLENGES DURING CANCER CARE

Across the world, breast cancer is the most common cancer among women, with approximately 1.7 million women diagnosed annually (Ferlay et al., 2015). In the United States, one in eight women are diagnosed with breast cancer in their lifetime (Siegel et al., 2015). Although treatments are continuing to improve, patients still face numerous challenges that impede their ability to access care or negatively affect their quality of life. Studies examining the cancer care system in the United States have identified numerous challenges individuals face during cancer treatment and survivorship related to managing medical treatments, emotional wellbeing, and treatment logistics (Salonen et al., 2011; Hayes et al., 2008; Unruh and Pratt, 2008a).

Patients must balance a variety of tasks to effectively manage their medical care. Such activities can include tracking the progress of their care, tracking side effects, and organizing information from multiple providers (Unruh and Pratt, 2008b). Health information seeking is one difficult task that individuals become immediately responsible for upon diagnosis. Finding relevant health information to understand one's disease and treatments can be critical for decision support and effective health management. Research has consistently shown that cancer survivors wish for more health information, particularly information that is specific to their diagnosis and treatment options (Beckjord et al., 2008; Jenkins et al., 2001). Obtaining this tailored information can be difficult. Providers have limited time with patients, and as previously mentioned patients do not always know what or who to ask. Although more people are turning to the Internet for health information (Fox and Purcell, 2010), online information sources are often not trusted and amount of information available can be overwhelming.

Coping with the emotional repercussions of a cancer diagnosis can be one of the greatest burdens that patients must overcome. Depression and anxiety are prevalent among cancer patients (Shapiro et al., 2001). The emotional consequences of a diagnosis persist even once an individual completes treatment. Studies have revealed the persistence of loneliness and fear of recurrence through survivorship (Cappiello et al., 2007; Rosedale, 2009). Thus, while managing the physical effects of cancer and treatments, patients must also deal with lasting emotional impacts.

In addition to managing one's physical and emotional health, patients often face a number of logistical challenges accessing care and obtaining health information. The cost of cancer care and health insurance in the United States are profound stressors for many individuals diagnosed with cancer. The high cost of treatment can have a severe impact on health outcomes and patients' quality of life. A recent study found that cancer patients were more likely to declare bankruptcy than individuals without cancer (Ramsey et al., 2013). Although health insurance can alleviate some of the expenses, individuals with cancer can be responsible for $25,000 (or more) annually on cancer treatment, making care unaffordable even for individuals with health insurance. Over 50 million Americans are uninsured or underinsured, amplifying the significance of treatment costs as a barrier to care (Kirkwood, 2016).

Thus, while improvements in medicine have significantly decreased the mortality rate of breast cancer, individuals going through breast cancer treatment continue to face a range of challenges. Many of the common challenges described previously highlight the importance of access to relevant health information and resources and can greatly impede on an individual's ability or willingness to receive care. In recent years, a body of literature has grown within the field of human–computer interaction, investigating how technologies can help patients manage such complex health care situations. Existing tools provide and organize health information (Tang et al., 2006), help users reflect on their experiences (Mamykina et al., 2008), and foster social support (Skeels et al., 2010), among other activities (Clauser et al., 2011). These technologies support individual tasks, but there is a lack of systems that offer continuity and connectivity between tasks, people, and content. In addition, rarely do we find tools that possess the flexibility or robustness to cater to an individual's needs, as they change over time. Owing to these gaps in support, the burden is placed on the patient to cobble together various tools that support their health care needs (Unruh and Pratt, 2008b). These tools often do not work in coordination with each other, adding to the patient workload as they must track and manage the disparate pieces of health care information and tasks. Furthermore, as circumstances change due to a new treatment regimen or a new health behavior or goal, the responsibility falls on the patient to find the resources that best suit their needs as they adapt to each "new normal". Thus, in our work we have collaborated with existing cancer care professionals to understand existing human practices for helping patients manage the range of challenges faced during the cancer journey and to develop tools to lessen the burden of personal health management during cancer treatment.

3. CASE STUDY: THE BREAST CANCER JOURNEY

Our work began with the goal of examining how cancer patients manage their health and specifically their use of key resources and support outside of traditional health care settings. Over the past 5 years, we have been partnering with the Harbin Clinic and Rome Cancer Navigators in Rome, Georgia to determine how mobile technologies can support patients' broad and changing needs.

Rome is located in the northwest region of the state of Georgia, in the United States. Two hospitals and one cancer clinic (Harbin Clinic, our health care partner) are located within Rome. Although the city has a population of 36,000, individuals from nearby states including Tennessee and Alabama will come to Harbin for treatment. Harbin Clinic offers a number of medical services including both medical and radiation oncology. The cancer care team (including both medical and radiation oncology) includes five doctors, a pharmacist, and a physician assistant. In addition to the Harbin Cancer Clinic, we worked closely with the Rome Cancer Navigators, an organization we describe in detail through our case study. Generally, cancer navigators are health care professionals who offer individual support to patients throughout cancer treatment. For the first 2 years of our research, the navigators existed as a nonprofit organization, funded through a variety of philanthropic mechanisms. In 2015, the navigators became employees within the local hospitals and cancer clinic, with the nurse navigator working within a local hospital (Floyd Medical Center) and the service navigators working within Harbin Clinic. As we describe in our case study below, we first worked with Harbin and the Rome Cancer Navigators to understand this regional cancer care system. We utilized our understanding of the cancer care system to identify opportunities for further technological support for patients.

3.1 Understanding Navigation Practices

In our work, early conversations with survivors and oncologists revealed that cancer navigators were critical stakeholders within the cancer care system and worked very closely with newly diagnosed patients. Cancer navigation, developed in 1990 to address socioeconomic health care disparities in Harlem (Freeman, 1993), helps patients navigate the health care system, remove barriers to accessing care, and aims to reduce national health care disparities

(Robinson-white et al., 2010). Today, many cancer navigation organizations are localized, focusing on connecting patients with useful information and resources. Within our work, partnering with and understanding navigation practices were a critical step in understanding the cancer care system. By studying the daily practices of cancer navigators, we were able to examine patients' medical and personal support needs, learn how patients' cancer journeys changed over time, and better understand the interactions that exist among patients, doctors, and navigators.

At the time of our study, the Rome Cancer Navigators consisted of five employees who worked with over 900 patients a year. The organization's central focus was to reduce barriers to care by answering medical questions, offering emotional support, and helping patients receive the necessary health insurance or other important resources. The organization divided tasks among two nurse navigators, two service navigators, and one social worker. The nurse navigators utilize their medical backgrounds (both are registered nurses) to offer patients the medical information and answer health-specific questions, eliminating gaps in patients' medical knowledge. The service navigators, who have prior experience in social work, focus on eliminating barriers to care by connecting patients to resources, such as health insurance, transportation, wigs, and food stamps. Finally the social worker provides patients and their families with counseling, thus offering an additional level of emotional support. This focus on medical needs as well as helping patients with the numerous emotional, financial, and logistical challenges that arise highlighted the comprehensive, holistic support needed by patients and offered through this organization.

Examining the role of cancer navigators in the broader health care system, we saw navigators as human agents of continuity and capacity. The organization linked multiple parts of the health care system, including patients, oncologists, surgeons, and case managers, while adding knowledge and facilitating access to the system. In Rome, Georgia, the navigators had access to the cancer clinic's electronic medical records and doctors' schedules, allowing them to meet patients at existing appointments, where they introduced patients to navigation services and offered continuing support. The navigators also constantly interacted with the oncology team, attending regular meetings to discuss new cases. This level of access allowed the navigators to prepare for meeting new patients and have a better understanding of patients' medical situations so that they could tailor resources and answer treatment questions. They also added to this system their expert knowledge in nursing and social work and provided an additional point of contact for patient education and support.

In an effort to understand the daily routines of cancer navigators, we interviewed all of the Rome cancer navigators over a 6-month period. Interview topics included navigator responsibilities, challenges encountered, and technology use. We employed an iterative inductive analysis to identify themes within the data. A central component of our work included organizing navigator roles and responsibilities around established cancer phases (Jacobs et al., 2014b), summarized in Fig. 2.1. We found that navigators are particularly focused on offering support to newly diagnosed patients before they begin treatment. During this time the nurse navigators will meet patients at their first oncology appointment, provide medical expertise and information, and refer patients to the service navigators. The service navigators will meet with patients to identify needs or barriers to care and help patients acquire the necessary resources, including insurance, medical supplies, transportation assistance, and counseling.

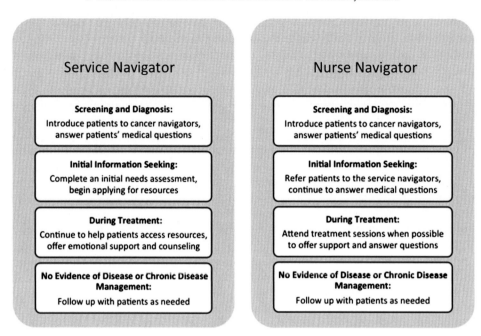

FIGURE 2.1   A summary of the cancer navigators' changing roles as a patient progresses through treatment.

As patients progress through treatment, the navigators will have less time as they must help newly diagnosed individuals but will follow up with patients as needed.

This work allowed us to identify the strengths and weaknesses of navigation practices and identify opportunities to amplify the human capabilities within the cancer care system. The following results of this work motivated our future research by providing guidelines for tools to support the cancer journey:

1. **Navigators offer consistency in a constantly changing process**: Patients can face surgery and several treatments, requiring them to constantly adjust to a new normal as they work with multiple doctors, attend treatments in various locations, and manage various side effects. During these difficult moments, the navigators pointed out the diversity of social support that exists among patients. As one nurse navigator stated, "*some people have a lot of support, others have no one.*"

   As previously mentioned, navigators have access to doctors' schedules (one nurse navigator stated that on an average week she will print 14 schedules each week). These schedules allowed the navigators to track their patients' treatments and thus make an effort to be physically present at key moments in the journey. Nurse navigators, in particular, spent much of their day at the cancer clinic. They met many patients at their initial oncology appointment, as well as initial chemotherapy and radiation appointments, when feelings of fear and uncertainty are common. Thus navigators offered consistent social support to patients through continuous physical presence at treatments.

2. **Navigators adapt to changing patient needs**: Examining navigation practices over time showed the extent to which navigators adapt their practices as patients move through the cancer journey, altering their roles and responsibilities as patients progressed through treatment. We found that typically nurse navigators will meet a patient at her or his first meeting with the oncologist and will introduce her or him to navigation services. Patients' early focus on gathering information about their diagnosis and treatment plans motivates this process. Nurse navigators served as their primary contact during this time. Many patients began to meet more often with a service navigator as they progress through the journey, as emotional support and minimizing socioeconomic barriers to care become the priority. Once a patient began treatment, the primary navigator was often the navigator the patient developed a close relationship with, and the responsibility fell on the patient to seek out navigation support as needed.

3. **Navigators personalize the care provided to each individual patient**: The service navigators in particular created a highly personalized process, guided by an initial patient assessment. This assessment, developed by one of the Rome service navigators, was filled out during the first one-on-one meeting with a patient. During this meeting, the navigator got to know the patient and began to match patients with the resources that best support their particular needs. The navigators have expert knowledge on the resources that are available to patients, such as transportation assistance, food stamps, counseling, and health insurance. By taking the time to get to know patients personally and connecting them with a tailored set of resources, the service navigators were able to help eliminate any barriers that could impede on a person's treatment.

4. **Large caseloads and scant information technology support limit the availability of long-term navigation**: The adaptive and personalized care offered by the cancer navigators provided continuity and flexibility within the cancer care system, benefiting hundreds of individuals managing cancer every year. Navigator services were particularly useful to newly diagnosed patients, as navigators concentrated on patients at the beginning of cancer journey due to the increase in medical questions and likelihood for barriers to significantly hinder care. During this time, nurse navigators were present at appointments and service navigators were able to connect patients to resources before beginning treatment. However, working with over 900 patients a year proved a significant hindrance for offering navigation services to those who further along in their cancer journey. After a patient began active treatment, navigators tried to occasionally follow up but faced time constraints. A navigator could have up to eight new patients in a single day, and thus their attention is often focused on the constant influx of new patients.

   Although information technology could help navigators monitor patients over the long-term and reach out to patients who are most in need of navigation services, no systems are currently in place to help sustain these patient–navigator relationships. In addition, navigators had minimal tools to share information with one another, manage their large caseloads, and propagate best practices. Therefore each navigator spent considerable time developing their own routines and relied heavily on their memory for recalling resources and case details.

In addition to the developing technology to support navigation practices, we saw an opportunity to build upon the navigation strengths described previously by developing tools that offer complementary personalization later in the cancer journey. Such tools could allow navigators to continue to focus on helping patients through the numerous challenges that occur early in the journey, offer continued support to cancer survivors, and help navigators monitor and connect with survivors when necessary.

3.2 Assessing Survivors' Journey Reflections

Our work with cancer navigators allowed us to study cancer care as a socio-technical system, determining key stakeholders and the interactions that occur between them. We also identified areas that lack technical support and began to find opportunities for technology to amplify the human capabilities and strengths within this system. Before focusing on the technical opportunities for supporting the cancer journey, we next worked directly with cancer survivors to understand their personal experiences and the challenges they faced throughout their journeys.

For this research, we worked with 31 cancer survivors, using both interviews and focus groups to encourage reflection of their health care experiences. Seventeen survivors participated in individual interviews, in which we used a semistructured method focused on participants' general health care experiences, as well as support and information needs. An additional 14 participants joined focus groups. Each focus group consisted of two to four participants and two researchers. We asked participants to write or draw their journey, focusing on moments that they felt significantly influenced their experience. We offered some general categories to encourage participants to think about their journey broadly, rather than solely on the medical perspective. Categories included significant moments involving family and friends, work and finance, moments of change, challenges they encountered, and emotional highs and lows. The focus groups proved to be a useful method for encouraging reflection, as survivors talked to each other and compared experiences, aiding participants' recollection of their journeys. The focus groups received positive feedback from participants, despite our initial concern of asking participants to talk about personal events with other cancer survivors.

Ultimately, through this work we developed a framework depicting the numerous responsibilities and challenges patients face and how these change over time (Jacobs et al., 2015a). A subset of the results are summarized in Fig. 2.2, demonstrating how a cancer patients' experiences change over time based on the reflections of our participants. Our framework offers guidance in the development of personal health tools that offer more holistic support to patients grappling with cancer. Some of the key results from this work are summarized in the following:

1. **Throughout the cancer journey, patients face a broader range of responsibilities and challenges than previously documented**: We identified 12 responsibilities and 11 challenges commonly discussed by our participants. We also assessed when, in their journeys, participants faced each responsibility and challenge. We used established cancer journey phases: screening and diagnosis, information seeking, acute care and treatment, and no evidence of disease (Hayes et al., 2008), as one axis in our framework. We then distributed responsibility and challenge themes across phases to show how

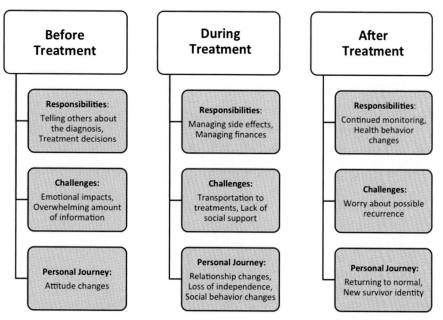

FIGURE 2.2 An example of the responsibilities, challenges, and personal impacts patients often experience during the cancer journey, as described by breast cancer survivors.

patients' priorities shift over time. For example, after diagnosis many participants shared that they were focused on telling others about their health situation, dealing with others' reactions to this news, and coping with the emotional impacts of the diagnosis. Once participants began treatment, they shifted to talking about the importance of managing symptoms, managing finances, coordinating transportation, and dealing with decreased social support. The range of responsibilities and challenges we identified shows that a need exists for flexible and adaptive support within personal health tools.

2. **In addition to responsibilities and challenges, patients grapple with a "personal journey"**: Through this work, we also identified a number of personal changes participants dealt with as they coped with having cancer in the context of their personal, daily lives. We call this the "personal journey": a part of the cancer journey that is unique to each individual and patient driven, as every participant defined cancer in her own way. For example, early in the journey many participants discussed dealing with internal and significant attitude changes. For some participants these changes meant recognizing their mortality, whereas others focused on living in the moment and appreciating life. Although many participants discussed attitude changes generally, the specific change was unique to each individual.

Cancer treatment triggered or motivated significant life events, from losing employment to embarking on a new college degree. Participants discussed the need to address these events as part of their cancer journey; for example, wanting to celebrate the completion of a college course alongside cancer treatments. These personal life events, though unrelated to cancer, became associated in participants' reflections of their cancer experiences, with no clear distinction between the two. The personal journey factors drive home the importance of supporting not only a patient's illness management but also more broadly the daily life and activities of the whole person.

3. **The cancer journey does not end with treatment**: Although we did not focus on changes that occurred throughout posttreatment survivorship, moving into this phase of the journey was undoubtedly a significant change for participants, bringing a new set of responsibilities, challenges, and personal changes. As participants completed treatment, they often had to cope with another new normal, as they visited the cancer clinic less often and developed new routines. Many participants pointed out that these new routines did involve close monitoring of their health to reduce the risk of a recurrence. We also found that many participants felt a strong desire to give back to the cancer community, either through volunteer work or by supporting new patients. These new goals and behaviors reveal that even if an individual completes treatment and shows no evidence of the disease physically, the impacts of the cancer diagnosis are lasting.

4. **Discrepancies exist between patients' and health care professionals' information sharing preferences**: By working with multiple stakeholders involved in breast cancer care, we were able to compare and contrast their views on the experience. Our work with navigators and patients, as well as additional interviews with providers, revealed discrepancies between the health information that patients were willing to share with health care professionals and the information doctors and navigators wished to receive (Jacobs et al., 2015b). Specifically many patients indicated that they were not inclined to share emotional impacts of the illness, such as feelings of loneliness, with their doctors or navigators. However, oncologists and cancer navigators stated that these were important factors to know in order to best support the patient. Patients explained that their reluctance was, in part, based on their (incorrect) assumption that health care providers were not interested in hearing about emotional challenges. We found a similar misalignment regarding satisfaction with care, as many patients said they would not feel comfortable sharing this information, while health care providers stated they would like to know when a patient is dissatisfied as early as possible to improve their care.

This second phase of research allowed us to focus on the patient perspective as it relates to the broader socio-technical cancer care system. The results of this work revealed the responsibilities, challenges, and personal influences of the disease that survivors found most significant. We used the results of this work, as well as our work with cancer navigators, to develop a flexible personal health technology, described below, to help patients cope with this broad range of issues.

3.3 My Journey Compass: Pilot Study of a Flexible and Mobile Personal Health Technology

In the first two phases of this work, we focused on understanding the cancer care system from the perspectives of health care professionals and patients. The research highlighted the dynamic nature of the cancer experience and the additional work placed on all stakeholders to adapt to these constant changes, as navigators adapt to patient needs and patients continuously cope with new transitions. We saw an opportunity to complement and expand upon the existing support, as navigators faced extreme time and resource limitations that prevented them from offering long-term support, yet the cancer survivors we worked with identified many challenges that occur during and after completing treatment.

In the final phase of this research, we utilized our previous findings to guide the design of a personal health management tool, used to assess how individuals with cancer utilize existing resources throughout treatment and survivorship. Specifically our goal was to provide newly diagnosed breast cancer patients with a suite of resources to support their personal and medical needs throughout the journey and assess how the use of the technology changed throughout participants' cancer journeys (Jacobs et al., 2014a, 2015c).

3.3.1 My Journey Compass Design

We began the development of the My Journey Compass system with three goals. First, we wanted to design a tool that was **flexible** enough to support participants' needs and goals throughout the entire cancer journey. We saw that flexibility was a critical component of navigation practices, as the navigators continually adapted their work to provide personalized care and support the different issues patients face. Furthermore, we did not expect a single application to be able to support the wide variety of patient needs. Thus we relied on a suite of tools and resource to provide the flexibility needed to support users throughout their cancer experiences.

Our second design goal was to create a **mobile** system. Prior research has shown that patients must complete a number of health management activities while away from their home or health care clinic (Klasnja et al., 2010). We decided to use a seven-inch Android tablet (the Nexus 7) to allow for easy portability and accessibility. In addition, the tablets provided an open platform that allowed participants to personalize their tablet experience, adding any applications they wished, adding additional flexibility to the system. Although we could have utilized smartphones, which are a more ubiquitous technology, we found that the tablets were small enough to be easily transported, while providing extra screen space to allow for larger text, a feature that was important to our participants.

Our final design goal for the My Journey Compass project was to develop a tool that was **integrated** with the existing health care system. As we describe below, we included health care providers early in the development process. This early partnership meant that the clinicians and cancer navigators understood and owned the technology intervention and could continue to support participants' use of the technology upon completion of the research project, as participants were able to keep the tablets after the study.

3.3.2 Health Care System Partnership

To create the final set of resources to be included on the My Journey Compass tablets, we worked with a team of oncologists, cancer navigators, breast cancer survivors, and directors from two local cancer centers. This team assembled a suite of PDF informational resources, applications, and websites that were considered useful to breast cancer patients throughout treatment and into posttreatment survivorship. Collaborating with these health care partners was particularly important for ensuring that participants received a suite of resources that were trusted and recommended by their health care team.

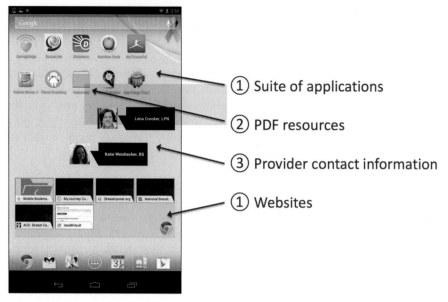

FIGURE 2.3   A screenshot of the My Journey Compass tablets. Participants were able to add their own applications to this preselected set of resources.

This design team included health information in PDF format or bookmarks to trusted Web pages, health applications, and personal applications on the tablets. Twenty-five PDFs and five bookmarks were included on the tablets. These documents presented a range of cancer-related information about topics such chemotherapy, radiation, lymphedema, breast reconstruction, exercising after surgery, fatigue, and more general guides for patients. Information pieces about cancer navigators were also included. Several mobile applications related to health were added to the tablets, including Paced Breathing, Cancer.net, Caring Bridge, My Fitness Pal, Nutrition Facts, and Relieve Stress. Finally, personal applications included participants' email and calendar. We were also able to include resources that were tailored to the participants' local health care system, including the contact information of their doctors and cancer navigators. Fig. 2.3 shows a screenshot of the My Journey Compass tablets.

3.3.3 Participant Recruitment

Our clinical partners were also involved in the deployment of the My Journey Compass tablets. Over the course of 1 year, all newly diagnosed breast cancer patients were invited to enroll in the study and receive a tablet. We recruited participants at their first appointment with an oncologist at the Harbin Cancer Clinic. The research team and cancer navigators jointly determined that cancer navigators would recruit participants into the study. This process had two significant benefits. First, introducing the technology and research through the cancer navigators allowed us to be sensitive to participants' needs during this emotional time, since navigators are experts at meeting patients during these difficult moments. Navigators could use their expertise to introduce patients to the technology in an appropriate manner, and we avoided overwhelming participants by asking them to meet the research team while coping with the new diagnosis. Second, recruiting participants through the navigators was beneficial as navigators are already present at patients' first meetings with the oncologist. Therefore we were able to incorporate recruitment into the cancer care system without disrupting existing processes.

3.3.4 *Creating an Education Navigator Position*

Although the cancer navigators were responsible for participant recruitment, we also wished to provide participants with training and technological support through the health care system. However, these tasks fell outside the navigators' daily responsibilities. Due to the fact that training would take place soon after a participant's diagnosis, the cancer navigators ultimately decided to create a new position, the education navigator, to accommodate these tasks. Throughout the study, the education navigator provided participants with a personal training session in which they reviewed the resources on the tablet and went over basic functionality of the tablet device. The education navigator also served as participants' point of contact when technical questions arose.

3.3.5 *My Journey Compass Deployment*

When a participant enrolled in the study, she or he received the tablet from the cancer navigators and was encouraged to use it any way he or she wished, with no restrictions. Participants were able to keep the technology and add any applications to the tablet, allowing us to assess how they choose to use the personal technology during their cancer care. Upon receiving the tablet, participants set up a training session with the education navigator, which typically occurred 1 to 2 weeks after the initial consultation.

Over the course of a year, we monitored the tablet usage of 36 participants. All participants were diagnosed with stage 0–III breast cancer at the Harbin Cancer Clinic. Of the 36 participants, 35 were female and ages ranged from 24 to 80 years (M = 60). At the study's conclusion, participants had possessed the tablets for a range of 170–365 days (M = 310). Throughout the year we interviewed participants and automatically logged the applications used and duration of use. We did not track application content such as search terms or social media posts.

Analyzing this data, we found that participants engaged with the technology throughout the cancer journey. On average, participants used the tablets for 2.6 h/week, and 14 participants continued to use the tablets after completing treatment and through the end of the research study. Our analysis of the interview data and the tablet usage revealed many insights into how technology may better support cancer journeys from diagnosis through posttreatment survivorship. We highlight some of these findings below.

1. **Mobility and privacy motivated adoption of the tablet**: Conversations with participants during the first month of the study revealed that participants initially began using the My Journey Compass tablets due to the mobility and privacy afforded by the technology. The mobility was particularly useful for participants who wanted to travel, as they felt more comfortable going out of town knowing they had their personal health information and doctors' contact information readily available. Many participants also shared that they brought the tablets to their treatments, particularly chemotherapy. During chemotherapy sessions, participants would use the tablets to play games, listen to music, or read books

as a way of staying calm and taking their minds off of the stressful situation. Participants also shared that the tablets easily allowed them to capture questions for their doctors while they were away from the clinic, and bring these questions to their next treatment. Participants noted that they felt more comfortable bringing the tablets to doctor appointments than the traditional binder of cancer-related information provided to patients after their diagnosis. Participants preferred to bring the tablets because, as one participant summarized, in the Rome, Georgia community people can easily identify cancer patients "because you have the big cancer folder". The tablets provided discretion that was preferred by many of our participants.

2. **The open platform allowed participants to extend the platform to meet their needs**: A unique aspect of this study was that we encouraged all participants to use the technology as their own, adding any applications or resources they wished. In contrast, most health care resources, whether digital or analog, are limited to clinical concerns and do not facilitate or encourage patients to find additional resources. Although we added a suite of health-related resources and information related to their diagnosis and treatment, participants could add any applications they wished, just as they would on commercial tablet computers. The majority of participants (35 of 36) did add applications to their tablets. Participants on average added nine applications to the tablet (ranging from 0 to 32), and, as a group, they added a combined 178 unique applications. Some commonly added applications included Amazon Kindle, the Bible, and Candy Crush.

   In reflection, we found that allowing customization led to two benefits. The first benefit was that participants could add resources that supported their personal needs and thus the tablets became more personally meaningful to each participant. For example, several participants added photos of their family members to the tablet, and therefore kept the tablets with them on a daily basis. Many participants (n = 13) also added religious applications. Adding these applications helped participants cope with their illness and thus supported their emotional wellbeing.

   The second benefit that came from providing participants with an open platform was that the research team was able to learn about participants' needs and values. The applications that participants chose to add to their tablets, and when they used these applications, revealed insights into how technology could better support their needs. For example, the increased use of games and other media during chemotherapy revealed the importance of supporting patients' emotional wellbeing during treatment sessions.

3. **Participants used nonmedical applications to meet health goals**: Since many cancer-specific tools were included in the tablet by the health care team, we found that participants did not add applications related to cancer, but they added many applications that would not be considered related to health care. Such nonhealth applications included religious applications and entertainment applications such as games, YouTube, and Facebook. In addition to making the tablets more entertaining and supporting participants' emotional wellbeing (as we described previously), these applications actually helped participants with their health-specific goals and activities. For example, participants discussed how they would use Facebook to share and receive health information with family and friends. Another participant explained that she

was beginning a vegetarian diet and used YouTube to learn new, healthy recipes. These examples illustrate the benefits that can come from developing personal health tools that are open and flexible, providing the capacity to meet a broad range of patients' needs and goals.

4. **Use of health information was episodic**: Through our assessment of tablet usage, we were able to identify a number of usage patterns. Four participants were power users, who used both the tablet and health information from the health care team regularly, with fewer than 2 weeks between each use. Another 12 participants were periodic users, who would use both the tablet and the health information on the tablet frequently for several weeks and then have long periods in which they did not use the tablets. The final and most common usage pattern, including 17 participants, involved continuous use of the tablet system and periodic use of the health resources. Subsequent interviews with participants showed that because the tablets were personally useful and meaningful to participants, they would keep the tablets with them and use them daily. While they opted to use the technology regularly, they would take purposeful breaks from using the tools related to cancer. Many participants described taking these breaks because they did not always want to feel like a patient nor did they want to feel like cancer was the focus of their lives. While they would not use the cancer tools regularly, by keeping the technology with them for personal reasons, they could easily return to using those resources when needed. Thus we found that by providing holistic support through the technology, we encouraged long-term engagement with the resources related to cancer.

5. **Cancer navigators were influenced by the deployment in several ways**: In addition to providing insight into existing cancer care practices, our partnership with the cancer navigators significantly helped the research project. Throughout the recruitment process, have the navigators introduce the technology to patients helped to demonstrate that the health care team supported the technology, encouraging participants' trust in the technology. We were also able to ensure that throughout the project, participants who had expertise in working with patients had a point of contact and could be sensitive to the difficulties participants faced.

   Although we expected the partnership with the cancer navigators to benefit the research, it is important that researchers are aware of how a study impacts the health care system. In this project, the most evident change to the health care system was the creation of the education navigator position. While this navigator led training sessions and technical support for the research project, the position also offered long-term benefits for the organization. Upon completion of the study, the education navigator was able to continue to offer useful support, helping the navigation organization to expand its own technology fluency and adding this technical expertise to the services offered to patients. We believe this type of position, that includes educating patients on the potentially useful technological tools and resources, could be a viable position that benefits other health care systems.

   We were also pleasantly surprised to find that the nurse and service navigators also saw benefits from the technology intervention. The navigators shared that they were able to use the study as an icebreaker when meeting with new patients. Offering participants a new tablet computer became a method for easing stress and introducing patients to navigation services.

4. LESSONS LEARNED: SUPPORTING HEALTH CARE JOURNEYS

In this chapter we have described three phases of research that have progressed our understanding of how personal health technologies may better support longitudinal and dynamic health journeys. First, we studied the existing socio-technical health care system through an analysis of cancer navigation work practices. Second, we determined how patients' cancer care experiences change over time and are integrated with nonmedical events and challenges. Finally, we used the My Journey Compass technology probe to examine how existing a suite of existing tools and resources can support patients' needs throughout the cancer journey. Here, we synthesize the results from this multiyear engagement, proposing a set of design guidelines to aid the development of personal health technologies for health care journeys. Although these guidelines are based on our work within the cancer care system, they offer useful advice for chronic illnesses more generally, which are regularly characterized by continuous activity and change.

4.1 Understand the Role of Privacy

A significant factor in the success of personal health technologies is the privacy it affords to users. Designers of health tools must consider how a new technological system provides or hinders a user's control over their data. Of course, issues such as security of personal data, accessibility, and a user's control over their personal data are all factors influencing a tool's privacy. However, basic design features such as the form factor itself can also greatly influence the privacy of the device. Ubiquitous mobile tools offer an approach for increasing the privacy afforded to users. In our case study, we found that participants felt that the tablets offered greater privacy than the traditional cancer binders. Individuals are able to use technology to access health information without revealing the activity to the people around them, a privacy not afforded by health systems' traditional paper-based information or closed, specialized health care devices.

4.2 Place Patients in Control

The responsibilities and challenges individuals face when managing a chronic illness are numerous and unique to each person. Furthermore, each individual will be affected by the diagnosis in his or her own way and will define their own set of goals related to their health care and personal life. Therefore each individual will require a unique configuration of tools to support these goals. For example, while some individuals will rely solely on the information given by their health care providers, others will look online or use social media to share and receive information with family and friends. Realistically, a single application will not be able to provide the comprehensive support that patients need. By creating tools that are open and flexible, we put patients in control of curating their own personal set of resources and support tools. Such flexibility allows technology to be personally meaningful to each individual and encourages long-term engagement with the technology.

4.3 Amplify Existing People Practices

The advantage of understanding existing health care socio-technical systems before a technology deployment is that you learn about the strengths of the existing people and processes. A need exists for designers of health care technologies to focus on identifying opportunities for personal health tools to amplify existing strengths of health care systems and expand the availability of such care. Furthermore, by looking beyond the patient–doctor relationship, we see that there are many more stakeholders who directly influence patients' health care experiences. Greater support of these individuals and organizations may not only improve the care available to patients but offer designers and researchers with valuable collaborations as well.

4.4 Provide Holistic Support: Supporting Life Goals Not Health care Goals

Various barriers exist that could impede on an individual's ability to receive care or properly self-manage their health, including responsibilities in daily life or a lack of transportation, health insurance, knowledge, or social support. The responsibilities and challenges related to one's health care are intermingled with those of daily life. Although health care journeys occur within people's daily lives, the journey does not define a person's life. Personal health technologies can best support patients by combining medical uses (reading health information, recording questions for doctors) with personal uses (using social media to interact with one's support network, storing photos of loved ones). We need to continue to challenge ourselves to consider how technology can better reflect and support an individual, and not just a patient. By doing so, we can actually increase the utility of health tools. By allowing participants to use personal health technology for purposes beyond their health care, we extend the support offered through the tool, encourage engagement, and allow users to return to using the medical resources when necessary.

4.5 Design for User's Changing Needs

We have yet to see the development of personal health tools that support patients' dynamic needs as they move through multiple phases of care. As individuals with chronic illnesses work to understand treatment options, prepare for upcoming treatments, and manage side effects, we must offer support that reaches across these different phases of care. Such tools could allow individuals to cope with existing challenges as well as prepare for future health care changes, which bring new uncertainties. Adaptive systems can help connect individuals with personalized tools that are tailored to their current situation, offering in the moment support that continuously updates with the most relevant resources. An opportunity also exists to abstract journey information for reuse as an individual progresses through the health care journey. As we continue to advance our understanding of chronic illness experiences, we may better utilize an individual's diagnosis and past behaviors to predict patient needs and develop targeted interventions.

5. FUTURE WORK

Our formative studies revealed a number of opportunities for utilizing technology to support the existing health care system, many of which we could not address in the My Journey Compass intervention. One interesting finding from our formative work was that the information that patients are willing to share with their providers does not always align with the information needs of their clinicians. This result requires more direct attention and highlights a need for systems that make information sharing more transparent. Such systems could help patients understand how health care providers can use the information they share to benefit them, while also helping providers better understand why patients are not comfortable sharing particular information.

6. CONCLUSION

Supporting health care practices requires an in-depth empirical understanding of the work and the broader system in which it exists. Thus, in our work, supporting patients' health management outside of the clinical setting, we utilized empirical studies that assessed the multifaceted nature of the cancer journey from multiple perspectives. First we worked with cancer navigators to understand how this existing organization offers personalized and holistic support to newly diagnosed patients. We then worked with cancer survivors to understand how their needs and goals changed throughout the cancer journey. This formative work led to the development of My Journey Compass, a tablet computer system that was flexible enough to support participants' needs and goals throughout the entire cancer journey. Our field study of patients' use of this system provided an initial validation of our approach of curating a suite of tools to help meet a diverse set of patient needs, integrating system deployment and use into the patients' health care system, and providing a mobile and open platform to facilitate a broader and sustained engagement by patients. These three pieces of research, taken together, offer guidelines for supporting patients' long-term health management better. Our hope is that technology designers can pull from these guidelines to inform novel approaches for enhancing the capacity and capabilities of the socio-technical systems that are increasingly important in supporting the health care journeys of people with chronic illnesses.

References

Abowd, G.D., Mynatt, E.D., 2000. Charting past, present, and future research in ubiquitous computing. ACM Transactions on Computer-Human Interaction 7 (1), 29–58.

Beckjord, E.B., Arora, N.K., McLaughlin, W., Oakley-Girvan, I., Hamilton, A.S., Hesse, B.W., 2008. Health-related information needs in a large and diverse sample of adult cancer survivors: implications for cancer care. Journal of Cancer Survivorship: Research and Practice 2 (3), 179–189. http://dx.doi.org/10.1007/s11764-008-0055-0.

Buntin, M.B., Jain, S.H., Blumenthal, D., 2010. Health information technology: laying the infrastructure for national health reform. Health Affairs 29 (6), 1214–1219. http://dx.doi.org/10.1377/hlthaff.2010.0503.

Cappiello, M., Cunningham, R.S., Tish Knobf, M., Erdos, D., 2007. Breast cancer survivors: information and support after treatment. Clinical Nursing Research 16 (4), 278–293. http://dx.doi.org/10.1177/1054773807306553. discussion 294–301.

Clauser, S.B., Wagner, E.H., Aiello Bowles, E.J., Tuzzio, L., Greene, S.M., 2011. Improving modern cancer care through information technology. American Journal of Preventive Medicine 40 (5 Suppl. 2), S198–S207. http://dx.doi.org/10.1016/j.amepre.2011.01.014. Elsevier Inc.

Ferlay, J., Soerjomataram, I., Dikshit, R., Eser, S., Mathers, C., Rebelo, M., Maxwell Parkin, D., Forman, D., Bray, F., 2015. Cancer incidence and mortality worldwide: sources, methods and major patterns in GLOBOCAN 2012. International Journal of Cancer. http://dx.doi.org/10.1002/ijc.29210.

Fox, S., Purcell, K., 2010. Chronic Disease and the Internet. Pew Internet & American Life Project. http://dx.doi.org/10.1071/PY04018.

Freeman, H.P., 1993. The impact of clinical trial protocols on patient care systems in a large city hospital. In: Interactive Workshop on Medical Ethics Versus Medical Eco- Nomics: a Health Care Dilemma in Cancer Patient Care, pp. 2834–2838.

Hayes, G., Abowd, G., Davis, J., Blount, M., Ebling, M., Mynatt, E.D., 2008. Opportunities for pervasive computing in chronic cancer care. Pervasive Computing 262–279.

Jacobs, M., Clawson, J., Mynatt, E., 2014a. My journey compass: a preliminary investigation of a mobile tool for cancer patients. In: Proceedings of the SIGCHI Conference on Human Factors in Computing Systems (CHI '14).

Jacobs, M., Clawson, J., Mynatt, E.D., 2014b. Cancer navigation: opportunities and challenges for facilitating the breast cancer journey. In: Proceedings of the ACM Conference on Computer Supported Cooperative Work, CSCW, pp. 1467–1478. http://dx.doi.org/10.1145/2531602.2531645.

Jacobs, M., Clawson, J., Mynatt, E.D., 2015a. A cancer journey framework: guiding the design of holistic health technology. In: Conference on Pervasive Healthcare (Pervasive Health '15).

Jacobs, M., Clawson, J., Mynatt, E.D., 2015b. Comparing health information sharing preferences of cancer patients, doctors, and navigators. In: Proceedings of the SIGCHI Conference on Computer Support Cooperative Work (CSCW '15), pp. 808–818. http://dx.doi.org/10.1145/2675133.2675252.

Jacobs, M., Clawson, J., Mynatt, E.D., 2015c. Lessons learned from a yearlong deployment of customizable breast cancer tablet computers. In: Proceedings of the Conference on Wireless Health (WH '15). http://dx.doi.org/10.1145/2811780.2811951.

Jenkins, V., Fallowfield, L., Saul, J., 2001. Information needs of patients with cancer: results from a large study in UK cancer centres. British Journal of Cancer 84 (1), 48–51. http://dx.doi.org/10.1054/bjoc.2000.1573.

Kirkwood, M.K., 2016. The state of cancer care in America, 2016: a report by the American Society of Clinical Oncology. American Society of Clinical Oncology 12 (4), 339–383. http://dx.doi.org/10.1200/JOP.2015.010462.

Klasnja, P., Hartzler, A., Powell, C., Phan, G., Pratt, W., 2010. Health weaver mobile: designing a mobile tool for managing personal health information during cancer care. In: Medical Informatics Association Annual Symposium Proceedings (AMIA '10)vol. 10, pp. 392–396.

Mamykina, L., Mynatt, E.D., Davidson, P.R., Greenblatt, D., 2008. MAHI: investigation of social scaffolding for reflective thinking in diabetes management. In: Proceedings of the SIGCHI Conference on Human Factors in Computing Systems (CHI '08), pp. 477–486.

Ramsey, S., Blough, D., Kirchhoff, A., Kreizenbeck, K., Snell, K., Newcomb, P., Hollingworth, W., et al., 2013. Washington state cancer patients found to be at greater risk for bankruptcy than people without a cancer diagnosis. Health Affairs 32 (6), 1143–1152. http://dx.doi.org/10.1377/hlthaff.2012.1263.

Robinson-white, S., Conroy, B., Slavish, K.H., 2010. Patient navigation in breast cancer. Cancer Nursing 33 (2), 127–140.

Rosedale, M., 2009. Survivor loneliness of women following breast cancer. Oncology Nursing Forum 36 (2), 175–183. http://dx.doi.org/10.1188/09.ONF.175-183.

Salonen, P., Kellokumpu-Lehtinen, P.-L., Tarkka, M.-T., Koivisto, A.-M., Kaunonen, M., 2011. Changes in quality of life in patients with breast cancer. Journal of Clinical Nursing 20 (1–2), 255–266. http://dx.doi.org/10.1111/j.1365-2702.2010.03422.x.

Shapiro, S.L., Maria Lopez, A., Schwartz, G.E., Bootzin, R., Figueredo, A.J., Jo Braden, C., Kurker, S.F., 2001. Quality of life and breast cancer: relationship to psychosocial variables. Journal of Clinical Psychology 57 (4), 501–519.

Siegel, R.L., Miller, K.D., Jemal, A., 2015. Cancer statistics, 2015. CA Cancer J Clin 65, 5–29. http://dx.doi.org/10.3322/caac.21254.

Skeels, M.M., Unruh, K.T., Powell, C., Pratt, W., January 2010. Catalyzing social support for breast cancer patients. In: Proceedings of the SIGCHI Conference on Human Factors in Computing Systems (CHI '10), pp. 173–182. http://dx.doi.org/10.1145/1753326.1753353.

Tang, P.C., Ash, J.S., Bates, D.W., Overhage, M.J., Sands, D.Z., 2006. Personal health records: definitions, benefits, and strategies for overcoming barriers to adoption. Journal of the American Medical Informatics Association 13 (2), 121–126. http://dx.doi.org/10.1197/jamia.M2025.records.

Unruh, K.T., Pratt, W., 2008a. Barriers to organizing information during cancer care: 'I Don't know how people do it.'. In: Proc. of AMIA '08, pp. 742–746.

Unruh, K.T., Pratt, W., 2008b. The invisible work of being a patient and implications for health care: '[the doctor is] my business partner in the most important business in my life, staying alive.'. In: EPIC '08, November, pp. 34–44. http://dx.doi.org/10.1111/j.1559-8918.2008.tb00093.x.

World Health Organization, 2005. Preventing Chronic Diseases: a Vital Investment: WHO Global Report.

3

Supporting Collaboration to Preserve the Quality of Life of Patients at Home—A Design Case Study

Khuloud Abou Amsha, Myriam Lewkowicz
Troyes University of Technology, Troyes, France

1. INTRODUCTION

In this chapter, we report on a case study investigating the collaborative practices of a group of French self-employed health care professionals.[1] These caregivers decided to form an association—named e-maison médicale—dedicated to taking care of patients at home. The association is very successful in the sense that it allows patients who want to stay at home with their family to do so. However, its members are facing a huge work overload and express frustration at being unable to share and extend their collaborative model to include more caregivers and more patients. They came to us, asking to study their situation and to help them designing a system that could support their practices to define a more sustainable model. Our objective was then to understand how these health professionals create and adjust their collaborative practices and how to design technology that supports this collaborative process. This study contributes to the existing literature that investigates collaboration in home care context and provides insights on the complexity of supporting the collaboration among self-employed health care professionals who do not belong to any organization and do not follow any predefined protocol nor use a common information system.

To achieve our objectives, we decided to follow the design case study approach (Wulf et al., 2011), that starts with an empirical analysis of the practices in a particular field of application, followed by the design of an Information and Communication Technologies (ICTs)

---

[1] By self-employed health care professionals we refer to all health professionals and care workers who adopt an independent practice, that is, they are in charge of the economics of their practice: general practitioners, nurses, speech-language pathologists, physiotherapists, occupational therapists, dieticians, etc.

artifact depending on the findings, an implementation of the artifact, and an investigation of its appropriation over an extended period of time.

The results of our empirical work show: (1) the centrality of the coordinative artifacts (e.g., a liaison notebook) for sharing information and coordinating the work. (2) How focusing on patients' quality of life leads caregivers to address issues beyond the medical scope. (3) How team members experience different rhythms of collaboration depending on the patient's situation. This empirical work also permitted to identify challenges related to integrating new caregivers, sustaining the ongoing negotiation of roles and tasks, and motivating the engagement of all caregivers.

Based on these results, and motivated by the goal to tackle the listed challenges, we defined some design principles to support this type of ensemble. Following these principles, we developed the CARE application (Classeur pour une Approche en Réseau Efficace), which is accessible via a tablet and designed to stay at the home of the patient. The CARE application represents the technological component of a socio-technical system (Leonardi, 2012) that we implemented to support collaboration in the case of e-maison médicale. The CARE application is accessible via a tablet that stays at the patient's home. The patients keep the tablet with them when going for a consultation. All the health care professionals working with a patient can use the application.[2] We deployed CARE in five households and followed how it was used (or not) during 5 months.

The rest of the chapter proceeds as following: first we review studies on supporting collaboration in the context of home care, and then we describe the case of e-maison médicale as well as some background information about home care in France. Next, we detail the method we adopted, and we report on our results. We describe the design principles coming from these results that we followed to build the CARE application and we report on the pilot study. Finally, we conclude with some lessons learned from this design case study.

2. BACKGROUND: SUPPORTING COLLABORATION IN HOME CARE

Home care has progressed to comprise not only the conventional health professionals but also the social workers who facilitate the patients' improvement and well-being. Nowadays, home care depends on care networks that involve informal caregivers (family members, friends, or neighbors), care workers, and professional caregivers (nurses, physiotherapists, dieticians). The broader network of caregivers might also include pharmacists and technicians (Consolvo et al., 2004). In addition to people involved on an individual basis, we can see institutions involved in providing home care, like community care centers, call centers, and providers of social and technical services (Bratteteig and Wagner, 2013).

Many studies explored the home care work as well as the actors involved in it. Some studies focused on the mobility of caregivers and its implication for their cooperation. In fact, the majority of caregivers are mobile and meet rarely, so it is difficult for them to achieve collaborative tasks like scheduling meetings, information distribution, information retrieval, short-term treatment coordination, and long-term treatment planning (Pinelle and Gutwin, 2002;

---

[2] The patient is the owner of the information on the tablet.

Nilsson and Hertzum, 2005). Caregivers spend little time in a shared office, which makes the chance to have opportunistic collaboration rare (Bricon-Souf et al., 2005), and formal collaboration may be challenging due to schedule variability within the team (Pinelle and Gutwin, 2002, 2003; Nilsson and Hertzum, 2005). Thus, some caregivers adopt a loosely coupled way of organizing collaboration to preserve their autonomy. In this mode of organization, caregivers minimize collaboration and interdependencies to deal with the unpredictability of the work setting (Olson and Teasley, 1996; Grinter et al., 1999). When they need to collaborate, caregivers attempt to initiate contact with others in ways that minimize the effort (Brown and O'Hara, 2003). For example, caregivers prefer asynchronous communication as it allows them to overcome uncertainty about others' schedules, locations, and availabilities (Pinelle and Gutwin, 2003; Bricon-Souf et al., 2005). However, caregivers still need tight coordination to accommodate the evolution of the care recipient's condition (improved or deteriorated; Nilsson and Hertzum, 2005).

Some studies focused on the use of artifacts to facilitate the collaboration between caregivers who are expected to coordinate within their organization (e.g., between work shifts) as well as across organizations. Thus, the caregivers are required to communicate and coordinate their activities across both their disciplines and their organizational boundaries (Petrakou, 2007, 2009). To meet the challenge, the caregivers involved might create tools and conventions to collaborate (e.g., SVOP binder [Petrakou, 2007], liasion notebook [Abou Amsha and Lewkowicz, 2014]). In addition to disseminating formal information (related to patient status of health, e.g., administrated medications), artifacts are also usually used to support the informal conversations that take place in home care work (Westerberg, 1999; Hardstone et al., 2004; Abou Amsha and Lewkowicz, 2014). We can find similar results in the work done on the use of medical records in hospitals where health professional use post-its on the official medical record to support informal conversations (Fitzpatrick, 2004).

The design case study we present in this chapter contributes to the existing literature that investigates collaboration in home care. Compared to other studies, our work focuses on self-employed health care professionals, who do not belong to any formal organization that would prescribe work and communication procedures, nor use a common information system. Their collective organization around the patient is self-regulated, and contrary to the usual practices of health care professionals (at least in France) they value equally the work of all of them regardless of their profession (general practitioner, specialist doctor, nurse, care worker, etc.). They even emphasize that doctors are not the ones who know the patients best, and that they rely on the information from the other care actors who spend more time with the patients.

3. THE CASE

3.1 Context: Home Care in France

The French health care system allows patients to freely choose their caregivers (Rodwin, 2003), and in France, home care services are mainly provided by self-employed health care professionals who have an independent practice (Chevreul et al., 2010). Since 2006, the reform of the Health Insurance encourages the creation of "coordinated care pathways" in which the general practitioner (GP) occupies a pivotal role. Chosen by the patient, the GP performs primary care and, if necessary, directs the patient to specialty care. The reform proposes better reimbursement (by Social Security) for the patients who follow the "coordinated care pathway" that has been defined for them. All the health care professionals involved in this care pathway (nurse, physiologist, care worker ...) can be chosen by the patient. In addition, patients have the right to change their GP and to address a specialist doctor directly.fn33[3]

The self-employed health care professionals work in isolation, but they have direct relation with the patients and their family members. Some patients require the intervention of multiple care professionals to stay at home. However, these different professionals working with the same patient do not always communicate or share information about the patient, which might affect the health condition of the patient and sometimes their safety at home. Usually, the patients and their family members are the ones who transmit information from one professional to another, and who organize the different visits. For example, a patient who suffers from a chronic condition requires having a close eye on the progress of her condition to avoid unnecessary acute accidents, but her situation is not so critical that she has to be sent to the hospital. To take care of this kind of complex situation, the self-employed health care professionals have to communicate with each other, and to coordinate their work.

Recently, more and more local innovative initiatives aiming at organizing the efforts of self-employed health care professionals around patients at home have emerged. This move is encouraged by the reforms of the French health care system and the need for new offers of home care services. In the following, we present the case of the e-maison médicale association, which represents one of these few successful initiatives for promoting collaboration in the domain of home care in France.

3.2 E-maison Médicale—A Local Initiative for Home Care

The e-maison médicale association gathers different self-employed health care professionals, located in the metropolitan area of Troyes (N–E of France). They aim at promoting a collaborative approach to home care delivery. The association was created in 2011. Currently, the association has about eighty members including various professions: care workers, physiotherapists, biologists, GPs, pharmacists, and nurses. Professionals involved in home care do not have any shared responsibility for the patient's situation. Each of them is responsible for his/her acts.[4] Depending on each patient's needs, the care group might include care workers, nurses, dieticians, pharmacists, specialists, GPs, and mental health services. A patient can benefit from this collaborative care if her GP (or any other care professional who treats her) is a member of the e-maison médicale association.

The members of e-maison médicale consider collaboration as essential to preserve the quality of life of the patient at home. This collaboration is complicated due to the members' overloaded schedules as well as the absence of a shared information space. In fact, the members of the association are skeptical about sharing information about their patients in their

---

[3] In this case, they will be reimbursed at a lower rate.

[4] The health care professionals do not sign a contract and they do not share liability.

"official" personal medical record (DMPfn55[5]). They claim that the DMP is an administrative system, useful for traceability and reimbursement of medical acts (by Social Security and insurance companies), but that it cannot support collaboration around the patient. In addition, the DMP is not accessible for all the care professionals, for instance care workers cannot access it.

The association aims at motivating all the actual patients' care professionals (who are not necessarily members of e-maison médicale) to collaborate, which is not obvious because care professionals in the primary sector are strongly attached to their independent practice. Therefore, e-maison médicale does not standardize the work practices but rather tries to combine different practices and skills to improve the quality of home care delivery.

4. METHOD

4.1 Data Collection

We collected data in mainly two situations: (1) when studying the actual practices of e-maison médicale, with ethnographic methods (Randall et al., 2007), combining interviews, observation, and discussion sessions, and (2) when discussing with e-maison médicale the design ideas that emerged from our findings, during two design workshops in which we used mockups and scenarios.

We conducted a field study over a period of 15 months in which we focused on the coordinative practices and artifacts (Schmidt and Wagner, 2004) of the e-maison médicale association. We started with a discussion session with five members of the association. Participants included a GP, a registered nurse (cofounders of the e-maison médicale), a physiotherapist, and two care workers. We recorded the discussion and noted down remarks. This discussion session lasted 3 h and motivated us to conduct an observation to see how actors coordinate their work in situ. Hence, we followed the registered nurse (one of the two founders of the association) for 3 days (15 h total). We visited 20 patients' homes per day. We took photos and noted down information.[6] During and after each visit, we asked questions to the various caregivers (mainly care workers and family caregivers).

This observation gave us a useful insight of the care practices of the members of e-maison médicale and highlighted the important role of the "liaison notebook" in the patient's home (Abou Amsha and Lewkowicz, 2014). The notebook provides an asynchronous way of sharing information and communicating among the different people involved in the care of the patient.

To go further, we organized a new discussion session with the founders of the network (a registered nurse and a GP), focusing on how the notebook supports the collaboration between the caregivers. The session lasted 2 h; we took notes and photos of the different liaison notebooks, we recorded the meeting and analyzed the transcript. We also collected a sample of eleven liaison notebooks (Table 3.1).

---

[5] In French, Dossier Médical Personnel (DMP) is the electronic medical record that has been chosen by the French ministry of health. It is secured and accessible via the Internet.

[6] Our main focus was the work of the nurse, and before taking photos we took the authorization of the patients or their family's members.

TABLE 3.1   The Collected Liaison Notebooks

Patient	Number of Notebooks	Number of Pages	Period of Care
SS	1	50	11/2011–06/2014
MD	1	84	08/2011–06/2014
LD	1	100	11/2011–05/2014
SG	8	340	04/2007–05/2014

The analysis of these empirical data (see next section) led us to some design ideas that we assessed during two design workshops that were also the occasion to collect data about the work practices of the members of e-maison médicale. The first design workshop lasted 4 h and a half, and we had six participants: three home-helpers, a registered nurse, a physiotherapist, and GP. We used mock-ups and scenarios. The three scenarios addressed the collaboration of regular caregivers and the participation of one-time caregivers in the collective management of patients at home. Participants had printed copies of the mockup, and they commented on our propositions and suggested new ideas. At the end of the workshop, all the ideas were arranged on a board. We filmed the workshop, took photos, and wrote notes.

The second workshop lasted 4 h, and we had six participants: three care workers, a registered nurse, a GP, and a specialist (all participants, except the specialist, had participated in the first design workshop). We presented a prototype of the application with use scenarios. Participants worked with the prototype and gave us valuable feedback. We were three researchers working with participants, taking notes, and photos.

During the 3 years of our research work, we also joined the monthly meeting of the association, where members discuss their practices and work on extending their logic of work to include more members. We were also involved on different occasions in some events concerning the e-maison médicale, like the presentation of their work to a new audience, or a meeting with the hospital of Troyes to define their collaboration. This involvement contributed to our understanding of their practices and the challenges they are facing.

4.2 Data Analysis Approach

We used open coding (Corbin and Strauss, 1990) to analyze the data collected from our different sources (interviews, observation notes, discussion sessions, design workshops). We iteratively coded the data; it took us three rounds of coding. The coding rounds were conducted by the same researcher but the resulted codes were discussed with a second researcher. In the first round, we were looking at the collaborative practices, and we defined codes like "acting together" or "asynchronous coordination". We also coded the different kinds of information the people were sharing; so defined codes like "medical instructions", "clinical finding" or "logistic needs". In the second round, we identified a relationship between the collaborative practices adopted by the care professionals and the kind of information they were sharing, thus we coded different situations where we found a pattern of practice-information structure: e.g., "treating an emergency", "modifying care plan", or "solving a problem". In the third and final round of coding, we identified a second level of classification related to different dimensions of the management of the patient's conditions: "medical", "logistic", and "social. This classification highlighted how care issues emerge, and how these issues that might span multiple dimensions are treated collectively.

5. RESULTS OF THE EMPIRICAL ANALYSIS OF PRACTICES

Obviously, managing the care of the patient at home differs from caring for the patient in a hospital. The home of the patient undertakes modifications in the place and lifestyle to enable caring for the patient safely. Receiving care at home is not only conditioned by the capacity of providing medical care at home, but also by the ability of the home of the patient (place and people) to afford the cost and the burden of care activities. To create a sustainable model for home care, the members of e-maison médicale extend their objectives beyond medical care to include maintaining the quality of life of patients and their families. In the following, we present how this model takes place (Abou Amsha and Lewkowicz, 2016); first we present the "liaison notebook" as a coordinative artifact. Then, we show how issues spanning the medical, logistic, and social dimensions challenge the provision of home care. Finally, we describe how caregivers experience different rhythms of collaboration to handle emerging issues.

5.1 The Liaison Notebook as a Coordinative Artifact

Home care professionals work mainly asynchronously; thus, the paper-based liaison notebook offers them a way to communicate about the situation of the patient. The practice of documenting information about the patient varies according to the conditions of the patients. The patients who suffer from chronic diseases, like diabetes, need more precise monitoring and therefore have specific notebooks designed for this kind of reporting. Some liaison notebooks might be less structured, and according to the evolution of the patient's conditions, structured medical information coexists with freestyle messages. Finally, some liaison notebooks represent a record of exchanged messages in a freestyle way. A message might include a mix of physiological measurements and observations, clinical findings and remarks about the patient's state of health. Thus, this style of documentation results in an ongoing, asynchronous, conversation between the care professionals about the patient's situation.

Most of the liaison notebooks we have observed include information about who are the care professionals working with the patient; usually there is a list of their names and contact information in the first page. This enables new care professionals, or one-time care professionals to contact the current care professionals if they need further information about the situation of the patient and her current care plan. We might also find in some of the notebooks a page describing elements of the patient's medical history, but sharing this kind of information is still problematic because not all of the care professionals officially have the right to read the patient's medical information (to ensure medical privacy). All liaison notebooks accommodate comments related to the management of the patient's care plan without being, in a strict sense, part of the medical information. This might be explained, we believe, by the need for organizing other aspects of care management to keep the patient safe at home. We develop in the next section the different aspects of care addressed by the care professionals.

5.2 Addressing the Multiple Dimensions of Home Care

Our empirical study highlights that providing "quality home care" requires dealing with issues beyond the medical scope. When creating a care plan, the care professionals handle the medical conditions of the patients, as well as their social situation, and even the home configuration. Indeed, they take into consideration whether the patients have any family caregiver, or whether they receive any financial and material support, or if they have enough room for medical equipment. Through our data analysis, we identified issues related to three facets of care: medical, social, and logistic issues.

Medical issues: To keep the patient safe at home, care professionals are challenged everyday by medical issues; care professionals collaborate to anticipate emergencies and to deal with problems properly. Medical challenges include keeping a patient stable, handling the potential secondary effects of the medication, and handling accidents that worsen the patient's condition. To manage the daily medical decisions, the care actors rely on the vigilance of each other. Patients with chronic diseases are an example where monitoring plays a significant role. The different care professionals meet rarely, and the absence of a shared history of the patient might affect the patient safety. Care actors handle current medical issues and anticipate possible future problems; to do so, they depend on their experience and acquaintance with the patient to perceive signs of worsening of the health status of the patient. Care actors share their views and insights on the notebook. However, if the care professionals see signs of potential risk for the patient, they call each other and try to fix the problem and avoid the emergency. When the problem requires some changes in the care plan, all the care professionals work together to stabilize the patient's situation.

Social issues: Keeping the patient at home safely depends, in many cases, on the involvement of informal caregivers.[7] Indeed, when patients are fragile (cognitively or physically, or both), the role of the informal caregiver becomes vital to ensure the safety of the patient. Thus, the care professionals look after the informal caregiver as an integrated part of the necessary efforts to sustain the home care for the patient. We see this when the care professionals use their competence and their relation to accelerate the caring process for family caregivers. Hence, the intervention of the care professionals is not limited to medical care; they are reactive to the modification happening in patients' social environment, and they reorganize the patients' care to ease the charge of the informal caregiver.

Logistic issues: Caring for patients at home include tasks like hiring care professionals, handling the medical equipment (functioning, maintenance), and modifying environmental safety hazards (like tripping obstacles, stairs without handrails). Logistic issues also include dealing with administrative formalities (ex. asking for prescriptions or medical appointments), as well as addressing daily issues related to medical equipment problems. Care actors discuss logistic aspects when starting or modifying a care plan; they ask questions like "do we need special medical equipment?" "Can we have the required medical equipment at home?" "Do we need additional professional caregivers?" "Can the patient afford paying for extra caregivers?" "Can the patient have financial help for home care services?" All care professionals might signal a logistic issue, and they all comment on the issue and propose a solution.

---

[7] Informal caregivers are persons who care for the patient without being paid for it, usually a spouse or children, but it also might include friends or neighbors.

5.3 Articulating Different Collaboration Rhythms

We have illustrated above that the care professionals face issues spanning medical, social, and logistic dimensions. Thus, they have to collaborate to be able to address different aspects of emerging issues and to accommodate the requirements of the evolving situation of the patient.

We have identified two interchanging phases: a "standard" coordination rhythm and an "intense" one:

In the "standard" phase, the patient's situation is relatively stable, the care professionals handle emerging problems individually according to their roles, and they coordinate their work conforming to care plan. The "intensive" phase starts when unexpected (medical or not) events arise and lead to a crisis that is challenging the current care plan. All the care professionals then collaborate in modifying the care plan to come back to "normal". Usually, the care professionals organize a "care meeting" at the patient's home to characterize the problem. This meeting consists of a discussion of the problem and of the different possible solutions. All the care actors, including the patient and her entourage, might participate in the discussion depending on the treated issues.

The rhythmic way of collaborating between the care professionals allows them to collaborate when it is necessary. Self-employed health care professionals put a special value on time because of their overloaded schedules. Having a classic team, with regular meetings, predefined agenda, and future objectives is not possible with the tight schedules of these care professionals.

5.4 Challenges of the Actual Practices

This reactive organization reassures the patients because they have the feeling that they can count on the collaboration of the different care professionals when a problem occurs. However, we identify challenges regarding the sustainability of this kind of organization:

1. **Integrating new care professionals**. New professionals constantly join the care ensemble. Current ones guide them toward their integration into the group. But due to their very busy schedule, they advise the new care professionals to look at the liaison notebook to understand the collaborative practices taking place around a patient. Unfortunately, the new care professionals lack the necessary time to fully review the notebook. It is then difficult for them to obtain a global vision of the patient's situation and to figure out how the collaboration occurs.

2. **Nurturing the ongoing role negotiation**. The roles of the health care professionals change according to the evolving situation of the patient. For instance, a GP who is usually at the center of the care organization might have secondary roles according to the addressed issue: solving problems related to the design of a bathroom to avoid falls, or to the difficulty of a patient to walk will not involve the same caregivers. In the collective approach of home care, the center is changing according to the nature of the emerging problems to be solved. While all the care professionals can participate in addressing the emerging issues, the leading ones change according to the addressed issue. This makes it difficult for the care professionals to find their place in this dynamic collective management. Thus, many of them focus on their individual tasks and watch the dynamic role negotiation "from outside".

3. **Ensuring the constant participation of all the care professionals**. The rhythmic collaboration affects the involvement and the motivation of the care professionals. While they participate actively into the intense collaborative episodes, it is difficult for them to keep the same quality of coordination during the "standard" collaborative phases. Having very busy schedules, they would not dedicate time for coordination activities if they did not see a direct benefit for the patient or their work overload. The situation becomes problematic when a patient is encountering a relatively long "standard" collaborative phase.

Acknowledging these challenges, and based on our analysis of the collaborative practices of the caregivers, we make the hypothesis that ICTs can offer the care professionals a way to visualize their collaboration, which, we assume, will enhance their motivation and facilitate the integration of new care professionals when needed. In the next section, we present the system that we have proposed, developed, and tested with the care professionals we have followed during our fieldwork.

6. DESIGN AND EVALUATION OF THE CARE APPLICATION

In this section, we present the design process of the (CARE—*Binder for an efficient networking approach*) application. First, we introduce the main design principles that we followed. Then, we explain how we translated these design principles into features. Finally, we report on the pilot study that we conducted for 20 weeks in the homes of the patients.

6.1 Design Principles

In the context that we have described above, supporting collaborative practices should allow different care professionals to participate in documenting the information concerning the patient. The care professionals need to discuss with each other about the condition of the patient, but due to their overloaded schedules, they rarely meet. Thus, we have to support their continuous discussions without disturbing their current workload. Finally, as the home of the patient is the place where the care takes place, we believe that the application should be made available at the home of the patient.

6.1.1 Enabling a Discussion-Based Documentation

Keeping track of the messages that have been exchanged between the care professionals and grouping the messages that address the same issues in a discussion thread provides a flexible way of documenting the information about the patient. First, it enables care professionals coming from different professions to explain their concern or request. In this way, the care professionals can not only document the facts about the patient's state of health but also explain or comment on what has been documented. Second, it provides a context to the documented information. Moreover, a discussion-based documentation is aligned with the current way the care professionals solve their problems and adjust their practices. Finally, a discussion-based documentation allows new members to have a look on how collaboration happened, and thus, it eases the participation of new care members in the discussion.

6.1.2 *Offering Tagging Possibility for Documented Information*

The open tagging allows care professionals to flag a part of a message (a physiological measure, a comment, a specific demand…) that they identify as important for the collective management of the care plan. This tagging offers a way to capture the elements emerging from the practice to help current actors highlight the important information that has to be considered when making decisions. Knowing that each patient offers a unique case, care professionals cannot predict what is the kind of information they will document or will need to achieve their work, and they cannot either predict what kind of issues they are going to address. An important and interesting aspect of this tagging solution is that it preserves the conversational context and the particular situation in which the information is collected.

6.1.3 *Tracking the Challenging Issues in a Patient's Trajectory*

Making the trajectory of the patient visible facilitates the integration of new care professionals by giving them the necessary information about the patient. In fact, the care plan reflects the current condition of the patient but does not offer the whole story. Tracking the challenging issues that arise can provide a vision of the case of the patient, and thus, allow the care professionals to understand the rationale behind the current care plan. To support this global vision, we suggest presenting a timeline in which the care professionals could mark the turning points in the situation of the patient. These marks can be annotated to explain changes in the care plan.

6.2 The CARE Application

In this section, we present the main features of the CARE application. The application is accessible via a tablet that stays at the patient's home. The care professionals participate in documenting the evolving caring plan. Regular care actors (including family members) can create a profile with their contact information, while nonregular care actors, for example a specialist doctor can access the application just by entering their name and profession.

6.2.1 *Enabling a Discussion-Based Documentation*

The CARE application offers a place where the care actors can exchange messages. Care actors can create a new message, comment on the existing messages, or they can acknowledge that they have read a message. When a care actor replies or comments on a message, a link appears at the bottom of the message indicating the name of the person who commented on it. Exchanged messages are presented in reverse chronological order, that is, the most recent message is shown first, we make the assumption that care actors read the messages frequently and that they are more interested in recent events.

All the messages that belong to the same thread can also be seen grouped together in the "discussion" page. Thus, caregivers can identify groups of threaded messages (comments and answers) to track issues that might trigger a change in the care plan. Each discussion is labeled with the title and the author of the first message and the number of messages it contains. The caregiver can click on the discussion to browse all the messages it contains in a chronological order.

6.2.2 *Offering Tagging Possibility for Documented Information*

The application allows the care actors to organize the information into categories in three different ways: First, they can store the information in a specific space, for example a list of medications, the patient profile, and care actors' profiles. Then, when creating a new message, the care actors can label the message as important; when a message is identified as important, it will be the first to appear in the thread regardless of its date of creation, until a care actor addresses the issue. Messages can also be labeled as "test results". A care professional can take a photo of a printed test result and comment on it in the message or she can simply indicate the result in a text message. The application does not aim to provide an archive for the medical tests but to offer a shared place for the information that is required by the care professionals to coordinate their activity, including some results of medical tests. Finally, the application allows care professionals to flag a part of a message either as an "alert" or a "physiological measurement". Our aim is to start with these two tags as a first step before providing a list of tags that could be created by the care professionals themselves.

6.2.3 *Tracking the Challenging Issues in a Patient's Trajectory*

The application offers a patient's profile that the care professionals can edit to add interesting information for the management of their patient. The patient's profile includes tables that group the information that was tagged in the messages of the care professionals. This collected information provides an idea of the condition of the patient; for example, the fact that the patient is falling frequently might signal deterioration. Thus, through the patient's profile, we can track the important events that affected or might affect the current care plan. These events are ordered chronologically, which offers a vision of the patient's trajectory. If necessary, the care professionals can track back the main message in which this event was tagged.

7. CARE PILOT STUDY

Adopting a summative perspective (Scriven, 1967), our main focus was to look whether the CARE application supports the collaboration between the care professionals and thus, contributes to the sustainability of their collaborative practices. The pilot study lasted 20 weeks (01/07/2015–30/11/2015). We equipped five households with tablets. The patients, their informal caregivers, and all their care professionals were allowed to use the tablet left at home.

7.1 Finding Candidates and Inclusion Criteria

We decided to include people with complex situations who were able to stay at home thanks to the intervention of multiple care professionals. We also tried to include patients with different care profiles. For example, we recruited two patients managed mainly by the members of e-maison médicale, three for which e-maison médicale members shared the management of the patient situation with other independent care professionals. So we finally included five patients: four proposed by GPs and one patient proposed by a registered nurse (Table 3.2).

TABLE 3.2   Patients Participating in the Pilot Study

Patient	Age	Number of Professional Caregivers	Number of Informal Caregivers	Comments
Mrs. SC	81	4	0	Only the nurse is a member of e-maison médicale
Mr. SS	73	6	0	Mainly taking care of by members of e-maison médicale
Mr. AA	75	6	5	Only the GP is part of the e-maison médicale
Mr. DR	80	4	1	Only the GP is part of the e-maison médicale
Mrs. KI	65	5	0	Completely managed by members of e-maison médicale

7.2 Rolling out CARE

The first time we went to the home of the patients and met patients and/or family members, we came with the tablet and a printed guide. The duration of our visit was between 60 and 90 min, during which we explained the objective of the pilot study, and we created together the different profiles which were going to access CARE on the tablet. All the participants signed an informed consent, indicating their agreement to participate in the pilot study.

During the same first visit, we tried to make a list of the care professionals of the patients to be able to contact them. We tried to fix an appointment with each of them for training them, usually at the home of the patient during their routine visits.

This second visit for training a care professional lasted between 20 and 30 min. During this visit, we showed the care professional how to create his/her profile and how to find information and write messages.

We also left a poster in each home that was indicating that the patient is participating in the pilot study and that all the care professionals who are taking care of him/her are invited to participate. We also left a paper-based guide explaining all the features of the application along with our contact information for any questions.

7.3 Follow-up and Data Collection

We collected data during and after the pilot study through regular visits at the patients' homes, and a discussion meeting with the involved care professionals at the end of the study.

The visits at the patients' homes were set up during the first visit: we fixed a weekly visit for the first 2 months. Once the patient and the care professionals were comfortable with the application, we reduced the visit to twice a month. During our visits, we checked if there was any technical problem or any questions about the application. We sometimes used the application together with the patient for writing a message for the other care professionals for instance. These follow-up visits lasted between 30 and 90 min each time. At these occasions, we frequently met care professionals doing their routine visits. They often had questions about the features of the application, and sometimes had some suggestions. These regular visits also offered us the opportunity to talk with new care professionals about CARE.

During these visits, we were taking notes, pictures, and a copy of the messages that were put into the application. The data were analyzed over the course of the study, which enabled us to ask more pertinent questions during our following visits.

We finally organized a discussion session with four of the care professionals who participated in the field study (a GP, a nurse, and two professional caregivers). They were members of e-maison médicale and had been using paper-based liaison notebooks at the homes of their patients for years. Two of them already participated into the workshops we organized during the design phase and were thus familiar with the application. The session lasted about 3 h and allowed us to get feedback about their experience when using the application. It also offered the opportunity for the different actors to discuss their views on the use of a device compared to the paper-based notebook.

We also picked some data collected from different patients' tablets to ask the professionals for some help in understanding the content. This session was video recorded, and we took notes and photos. This discussion session shaped our analysis of the whole data collected during this pilot study.

8. LESSONS LEARNED

We identified three topics related to supporting collaboration in this context, which are: (1) ensuring flexibility to accommodate different values, (2) building trust, and (3) open sharing. These topics are obviously interrelated: the flexibility allows the participation of a wild range of actors, which increases the chances of detecting issues and addressing complex issues collectively. The participation enabled by the flexibility facilitates creating a certain level of trust that is required for open sharing. Finally, open sharing allows caregivers to identify issues that might trigger intensive collaboration.

8.1 Flexibility to Accommodate Different Values

The care professionals believe that their collaboration is necessary when they want to keep a patient with a complex situation at home. However, they have different perceptions of the effort that is necessary to achieve this collective management of care. This idea can be illustrated by looking at the different perceptions of time among the care professionals: 5 min might be perceived as a short period for a care worker or a family member, but it might represent a full visit for a nurse.

This different perception of time is reflected in the diverging opinions about the efficacy of using the CARE application. Consequently, the health professionals like the nurse and the GP found the application more difficult to deal with than a paper-based notebook. On the contrary, home-helpers were much more positive about CARE. They were aware of the time needed to learn how to use the application, but they stated that once they were familiar with it, the application gave them more visibility on what was going on around the patient.

In summary, acknowledging the different values and perceptions of the different care professionals is the key to ease their participation. The collective management of the patient occurs thanks to the care professionals, thus, ignoring that they have different perceptions of some notions like time might affect their motivation and hinder their collaboration. A system supporting collaboration among a set of care professionals has to offer them a way to scan the participations of other care professionals in a short time and should help them to identify when there is something that needs their attention (like an alert) or intervention (like a question or demand). The difficulty resides in creating systems that adapt to the different potential users.

8.2 Building Trust

Our work with the patients highlighted the central role of trust when it comes to home care. This trust comes from different sources, like for instance the fact that the care professional belongs to a respected institution (this is especially the case for home-helpers), or the respect of the skills of doctors or nurses. Another source might be to trust a care professional based on a friend or a family member recommendation. It is rare to keep a caregiver when his/her work or attitude is not satisfying. This might be a particularity of the French primary care sector but it is important to mention it to understand how issues such as sharing medical information occur in this context.

The current care professionals played an important role in introducing us to the different patients. Thanks to the trusting relationship that existed between the patient and the different care professionals, we were accepted as an extension of the care process.

However, using CARE as a tool for sharing information between all the care actors (including the patient and the family members) was problematic, particularly for elderly patients. People were anxious about the introduction of technology that they do not control, and this was also true for some of the professionals. For example, one of the GPs was skeptical about participating in the pilot study because he thought it was illegal to write about the patient's medical situation. After we explained to him that the information is stored locally on the tablet of the patient, he agreed to participate in the experiment and signed the informed consent.

In the collective management of the patient's situation, the different care professionals share, though not officially, the responsibility for the patient. Thus, caregivers trust each other to start this voluntary collaboration. They share information and delegate tasks and count on each other's support when there is a problem.

This is reflected in the different ways the care professionals used CARE. For some patients, like Mr. AA and Mr. SS, CARE was used to facilitate the collaboration. The caregivers exchanged messages and addressed issues using the application. Most of the new care professionals created their profiles and started to participate in the discussions. The caregivers who did not really meet before were able to be introduced to each other and to exchange messages about the patient through the application.

However, for other patients, the application was used only to keep basic information like in the case of Mrs. KI where the care professionals organized their work but avoided documenting information on the tablet or on a notebook to protect the patient's privacy as she had problems with family members. In the case of Mrs. SC., the GP was completely absent from the application due the issue of trust as he could not control who can access to the medical information.

In summary, we argue that trust plays a major role in organizing work and collaboration in home care. We suggest that an application supporting collaboration between care professionals who meet rarely has to enable trust building between the different care professionals. In our case, the application participated in introducing different care professionals to each other and offered a place to start discussion between the caregivers. We believe that this provided the first step toward building trust and extending current collaboration.

8.3 Open Sharing

Care actors need to share medical and nonmedical information about the patient, which might be problematic as not all the care professionals are allowed to read medical information. This situation raised a lot of questions and discussions about the viability of technological solutions that offer open sharing to facilitate collaboration in the medical context.

Health professionals were skeptical about writing information related to the patient's medical situation on the tablet. In fact, the care professionals, particularly the members of e-maison médicale, already share medical information on the paper-based liaison notebooks. They consider this information as a "shared secrecy". According to the article N° L1110-4 of the French public health code the "shared secrecy" is made available either for health professionals to ensure the continuity of health care or inside institutions where the patient is taken care of by a team. Thus, the care professionals we met extend the notion of the shared secrecy because they trust each other, and they feel able to control the diffusion of the information by using the paper-based liaison notebooks. Here, we have to mention that the public health code is more explicit about sharing medical information through electronic transmission.

> To ensure the confidentiality of medical information […], the storage of this information in computerized formats, as well as their electronic transmission between professionals, is subject to rules established by decree of the State Council issued after public notice and the Commission's reasoned national data Processing and liberties. This decree determines where the use of the health professional card […] or equivalent device […] is mandatory. The health professional card and approved equivalent devices are used by health professionals, health care facilities, health networks, or any other body involved in prevention and care. article N° L1110-4 of the public health code.

Despite the restrictions of the regulation, the care professionals adopted CARE as an augmented version of the liaison notebook. The fact that the information is stored locally on the tablet of the patient has played a role in their acceptance. In our case patients own their medical information and they have the choice to share it (or not) with the care professionals.

When opting for open sharing, we made the assumption that the care professionals knew the information that was possible to share. For example, though the application offers a place to add the current medication of the patient, it is left to the GP to decide if it is necessary to fill it in. This is completely different from a medical Information System in which medication would be automatically added to the list when prescribed or when the medicine is bought at the pharmacy.

However, some care professionals mentioned that predefined categories of information might indicate what kind of information should be documented and thus enhance the usefulness of such application.

In summary, open sharing is required when care professionals collaborate in home care context because care professionals have to be aware of each other's views to identify issues that require to be addressed collectively. However, open sharing in the context of home care raises the questions of the reliability and the confidentiality of shared information.

8.4 Notes on the Implementation and the Training

About 60% of the 45 regular care professionals adopted the application and integrated it into their practices. Over the time, people's involvement increased. However, some features of CARE were not used, and the main features that were used were the creation of a new profile and the exchange of messages.

During the design workshops, the care professionals were pushing to get more features to categorize the collected information. For example, the list of medications was in the first mock-up a simple list in which care professionals enter the name of the medication and the dosage, but after two workshops, the list of medications became a complex form in which the care professionals have to indicate six different pieces of information before being able to add a new medicine into the list. Through the pilot study, we have noted that the care professionals were not using most of the features they asked for and advocated through the design workshops. When using the system, some of the care professionals found it difficult to understand how features like tagging could provide an answer to their needs, even though they participated in the design process and in collective presentations. Moreover, during the discussion session organized after the pilot study, the nurse and the GP who were the initiators of our collaboration were suggesting a feature to enhance the identification of important information and to be able to browse through the data, even though these elements were already supported by the CARE application (following their suggestions during the design workshops). While we avoided defending the application, the two home-helpers were actively demonstrating that these features already exist in the application.

Thus, surprisingly, the participation of the care professionals in the design process, although insightful, had little if any effect on the way they appropriated the application. In fact, the nurse and the GP who were very active in the two design workshops, and by then contributed to shape the application features, were the most critical of those same features during the pilot. However, the care professionals who had the chance to spend more time using the application better identified the different possibilities that the application was offering.

9. CONCLUSION

The design case study we presented in this chapter was set out to investigate collaboration that occurs between care professionals who collectively organize the care of patients in their home. Based on our findings, we suggested design principles that we implemented in our proposed socio-technical solution. We used the CARE application as a "technology probe"

(Hutchinson et al., 2003) to gain insights into how the work is done and how technologies might support the collaborative practices. The pilot study we conducted during 20 weeks thus allowed us to better understand the complexity of keeping the patient safe at home. It gave us further insights on how to design technology to support collaboration in the home care context.

To go further, this work could also lead to some implications in health policy; indeed, our fieldwork highlighted the importance of sharing information among the different care professionals to ensure the quality of care. Designing technologies to support collaboration in home care is hindered by a lack of adequate policy for sharing information. We claim that extending the notion of "shared secrecy"[8] might be a first step. We also propose that the patients or their informal caregivers should be able to identify who is involved in the home care and thus, who should have the right to access the shared information.

Finally, future research could focus on how to enhance trust building through communication. We think that the literature on social network at work offers an interesting start. Related questions include what motivates the use of social network at work (DiMicco et al., 2008) and the exploration of the different attitudes toward sharing information in such a context (Muller et al., 2010).

References

Abou Amsha, K., Lewkowicz, M., 2014. Observing the work practices of an inter-professional home care team: supporting a dynamic approach for quality home care delivery. In: Rossitto, C., Ciolfi, L., Martin, D., Conein, B. (Eds.), Proceedings of the 11th International Conference on the Design of Cooperative Systems (COOP 2014). Springer International Publishing, pp. 243–258. http://dx.doi.org/10.1007/978-3-319-06498-7_15.

Abou Amsha, K., Lewkowicz, M., 2016. In: Shifting Patterns in Home Care Work – Supporting Collaboration Among Self-employed Care Actors. Springer International Publishing, Trento, Italy.

Bratteteig, T., Wagner, I., 2013. Moving healthcare to the home: the work to make homecare work. In: Bertelsen, O.W., Ciolfi, L., Grasso, M.A., Papadopoulos, G.A. (Eds.), ECSCW 2013: Proceedings of the 13th European Conference on Computer Supported Cooperative Work, 21–25 September 2013, Paphos, Cyprus. Springer, London, pp. 143–162. http://dx.doi.org/10.1007/978-1-4471-5346-7_8.

Bricon-Souf, N., Anceaux, F., Bennani, N., Dufresne, E., Watbled, L., 2005. A distributed coordination platform for home care: analysis, framework and prototype. International Journal of Medical Informatics, Supporting Communication in Health Care Supporting Communication in Health Care 74 (10), 809–825. http://dx.doi.org/10.1016/j.ijmedinf.2005.03.020.

Brown, B., O'Hara, K., 2003. Place as a practical concern of mobile workers. Environment and Planning A 35 (9), 1565–1587. http://dx.doi.org/10.1068/a34231.

Chevreul, K., Durand-Zaleski, I., Bahrami, S.B., Hernández-Quevedo, C., Mladovsky, P., 2010. France: health system review. Health Systems in Transition 12 (6), 1–291 xxi–xxii.

Consolvo, S., Roessler, P., Shelton, B.E., LaMarca, A., Schilit, B., Bly, S., 2004. Technology for care networks of elders. IEEE Pervasive Computing 3 (2), 22–29. http://dx.doi.org/10.1109/MPRV.2004.1316814.

Corbin, J.M., Strauss, A., 1990. Grounded theory research: procedures, canons, and evaluative criteria. Qualitative Sociology 13 (1), 3–21. http://dx.doi.org/10.1007/BF00988593.

DiMicco, J., Millen, D.R., Geyer, W., Dugan, C., Brownholtz, B., Muller, M., 2008. Motivations for social networking at work. In: Proceedings of the 2008 ACM Conference on Computer Supported Cooperative Work, 711–720. CSCW '08. ACM, New York, NY, USA. http://dx.doi.org/10.1145/1460563.1460674.

---

[8] The notion is proposed according to the article N° L1110-4 of the French public health code and we discussed the notion in Section 8.3.

Fitzpatrick, G., 2004. Integrated care and the working record. Health Informatics Journal 10 (4), 291–302. http://dx.doi.org/10.1177/1460458204048507.

Grinter, R.E., Herbsleb, J.D., Perry, D.E., 1999. The Geography of coordination: dealing with distance in R&D work. In: Proceedings of the International ACM SIGGROUP Conference on Supporting Group Work, 306–315. GROUP '99. ACM, New York, NY, USA. http://dx.doi.org/10.1145/320297.320333.

Hardstone, G., Hartswood, M., Procter, R., Slack, R., Voss, A., Rees, G., 2004. Supporting informality: team working and integrated care records. In: Proceedings of the 2004 ACM Conference on Computer Supported Cooperative Work, 142–151. CSCW '04. ACM, New York, NY, USA. http://dx.doi.org/10.1145/1031607.1031632.

Hutchinson, H., Mackay, W., Westerlund, B., Bederson, B.B., Druin, A., Plaisant, C., Beaudouin-Lafon, M., et al., 2003. Technology probes: inspiring design for and with families. In: Proceedings of the SIGCHI Conference on Human Factors in Computing Systems, 17–24. CHI '03. ACM, New York, NY, USA. http://dx.doi.org/10.1145/642611.642616.

Leonardi, P.M., 2012. Materiality, Sociomateriality, and Socio-Technical Systems: What do These Terms Mean? How are They Related? Do we Need Them? SSRN Scholarly Paper ID 2129878 Social Science Research Network, Rochester, NY. https://papers.ssrn.com/abstract=2129878.

Muller, M., Sadat Shami, N., Millen, D.R., Feinberg, J., 2010. We are all Lurkers: consuming behaviors among authors and readers in an enterprise file-sharing service. In: Proceedings of the 16th ACM International Conference on Supporting Group Work, 201–210. GROUP '10. ACM, New York, NY, USA. http://dx.doi.org/10.1145/1880071.1880106.

Nilsson, M., Hertzum, M., 2005. Negotiated rhythms of mobile work: time, place, and work schedules. In: Proceedings of the 2005 International ACM SIGGROUP Conference on Supporting Group Work, 148–157. GROUP '05. ACM, New York, NY, USA. http://dx.doi.org/10.1145/1099203.1099233.

Olson, J.S., Teasley, S., 1996. Groupware in the wild: lessons learned from a year of virtual collocation. In: Proceedings of the 1996 ACM Conference on Computer Supported Cooperative Work, 419–427. CSCW '96. ACM, New York, NY, USA. http://dx.doi.org/10.1145/240080.240353.

Petrakou, A., 2007. Exploring cooperation through a binder: a context for it tools in elderly care at home. In: Bannon, L.J., Wagner, I., Gutwin, C., Richard, Harper, H.R., Schmidt, K. (Eds.), ECSCW 2007. Springer, London, pp. 271–290. http://dx.doi.org/10.1007/978-1-84800-031-5_15.

Petrakou, A., 2009. Integrated care in the daily work: coordination beyond organisational boundaries. International Journal of Integrated Care. 9 (3). https://www.ijic.org/index.php/ijic/article/view/URN%3ANBN%3ANL%3AUI%3A10-1-100567.

Pinelle, D., Gutwin, C., 2002. Supporting collaboration in multidisciplinary home care teams. In: Proceedings of the AMIA Symposium, pp. 617–621.

Pinelle, D., Gutwin, C., 2003. Designing for loose coupling in mobile groups. In: International Conference on Supporting Group Work, pp. 75–84. http://dx.doi.org/10.1145/958160.958173.

Randall, D., Harper, R., Rouncefield, M., 2007. Fieldwork for Design: Theory and Practice, 2007 ed. Springer, London.

Rodwin, V.G., 2003. The health care system under French national health insurance: lessons for health reform in the United States. American Journal of Public Health 93 (1), 31–37.

Schmidt, K., Wagner, I., 2004. Ordering systems: coordinative practices and artifacts in architectural design and planning. Computer Supported Cooperative Work 13 (5–6), 349–408. http://dx.doi.org/10.1007/s10606-004-5059-3.

Scriven, M., 1967. The methodology of evaluation. In: Tyler, R., Gagne, R., Scriven, M. (Eds.), Perspectives on Curriculum Evaluation. Rand McNally, AERA monograph series – curriculum evaluation, Chicago.

Westerberg, K., 1999. Collaborative networks among female middle managers in a hierarchical organization. Computer Supported Cooperative Work 8 (1–2), 95–114. http://dx.doi.org/10.1023/A:1008659328558.

Wulf, V., Rohde, M., Pipek, V., Stevens, G., 2011. Engaging with practices: design case studies as a research framework in CSCW. In: Proceedings of the ACM 2011 Conference on Computer Supported Cooperative Work, 505–512. CSCW '11. ACM, New York, NY, USA. http://dx.doi.org/10.1145/1958824.1958902.

4

A Community Health Orientation for Wellness Technology Design & Delivery

Andrea G. Parker, Herman Saksono, Jessica A. Hoffman,
Carmen Castaneda-Sceppa
Northeastern University, Boston, MA, United States

1. INTRODUCTION

Technological innovations are disrupting traditional notions of health care and wellness promotion. Personal health informatics (PHI) systems—applications that people interact with to manage their health—present an enormous opportunity to help lay people take an engaged role in their health care and overall wellness. Researchers have studied how such technologies can work in tandem with formal health care settings (e.g., during hospital stays and clinical therapy (Bers et al., 2001; Matthews and Doherty, 2011; Miller et al., 2016; Skeels and Tan, 2010; Wilcox et al., 2010)) and in everyday contexts that are separate from any institutional setting (e.g., health self-monitoring tools and games (Chen et al., 2014; Consolvo et al., 2008; Cordeiro et al., 2015; El-Nasr et al., 2015; Kay et al., 2012)).

Less research has explored the socio-technical context of designing and delivering PHI systems through community-based organizations (CBOs)—entities that provide services to geographically focused regions (e.g., neighborhoods), such as community centers, libraries, and churches. (Unless otherwise noted, we use the term *community* to refer to groups of people living in a constrained geographic area (Israel et al., 1998), such as a neighborhood or cluster of neighborhoods.) Yet, within public health, CBOs have been extensively employed as sites for health intervention design and delivery (Glanz et al., 2008). CBOs offer excellent opportunities for health promotion as they are often trusted entities within communities and provide an entry point for reaching potential users (Centers for Disease Control and Prevention, 2011; Minkler, 2005; Minkler and Wallerstein, 2008). They are particularly useful venues for reaching vulnerable populations (e.g., low socioeconomic status [SES] and rural populations) who are less-served by formal health care institutions, due to barriers such as cost and distance from medical facilities (Wright et al., 2013). At the same time, researchers confront new challenges when designing systems for CBOs as the values and practices of multiple stakeholder groups are intertwined (Minkler, 2005; Unertl et al., 2015).

This chapter examines the value of designing PHI tools that work in tandem with CBOs, as well as the challenges and considerations that arise in doing so. We present the results of a project in which we designed and evaluated a digital game for family exercise promotion (an *exergame*). This game was designed to be used within Family Gym, an existing public health intervention housed at community centers in Boston, MA, USA. Family Gym encourages physical activity among families living in low-income urban neighborhoods (Agrawal et al., 2012; Castaneda-Sceppa et al., 2014). Because of the extensive infrastructure of Family Gym and how tightly coupled it is to the community center in which it operates, we conceptualize Family Gym as a CBO in its own right. Indeed, we use *CBO* as an umbrella term to describe neighborhood institutions that serve local communities, as well as initiatives that are delivered within these communities—initiatives characterized by dedicated staff, and the delivery of regular programming and services to meet community needs.

Our work employs a socio-technical lens that focuses on the intertwined social, organizational, and technical influences on how PHI systems are designed and adopted, user attitudes toward these systems, and the impact that these tools have on individuals and communities. In this chapter, we specifically discuss the importance of a *socio-ecological* approach to PHI system design in a community-based organizational context. Socio-ecological health research involves designing interventions that account for the dynamic and bidirectional relationships of people and environments (Stokols, 1996). Taking this ecological perspective, our case study highlights how PHI tools can encourage wellness within a CBO context by addressing: (1) *intra-family* health-related values and practices and how they *align with and diverge from the CBO's values*, and (2) the opportunities and challenges that the *physical and operational facets of a CBO* provide for designing and disseminating family-based health technologies, particularly among underserved populations.

In the following sections of this chapter, we briefly review prior PHI research in community-based settings, overview our case study, and discuss findings from this research. We conclude by discussing the value of a community-based PHI research agenda focused on health equity, and opportunities for translating community-based participatory research (CBPR) models into PHI design directions.

2. PHI RESEARCH IN COMMUNITY-BASED ORGANIZATIONS

Public health interventions have extensively leveraged CBOs as sites for health and wellness interventions (Glanz et al., 2008). CBPR that equitably involves neighborhood organizations and leaders can help researchers develop culturally relevant interventions and delivery mechanisms (Minkler, 2005). Recreation centers, churches, schools, and many other institutional settings have been the sites of programs encouraging a range of healthy behaviors, such as nutritious eating and physical activity (Glanz et al., 2008).

As researchers have increasingly explored how PHI systems can effectively support wellness, most of this work has focused on formal health care environments (e.g., hospitals and clinics) or outside of any organizational settings (Bers and Cantrell, 2012; Epstein et al., 2014; Kay et al., 2012; Matthews and Doherty, 2011; Miller et al., 2016; Wilcox et al., 2010). Addressing the largely untapped potential of CBOs, public health researchers have begun to create PHI interventions that are delivered in neighborhood settings (Nollen et al., 2013; Smith et al., 2014). By offering these interventions outside of formal health care settings, CBOs can provide much-needed supports for low-SES populations with limited or no access to traditional health care and more sustained resources for prevention and chronic disease management (Wright et al., 2013). Furthermore, engaging with community organizations can help researchers create social systems for populations "affiliated by geographic proximity, special interests, or similar situations with respect to issues affecting their well-being" (Centers for Disease Control and Prevention, 2011; Nollen et al., 2013).

While the health sciences literature has demonstrated the impact of these interventions on health outcomes (Smith et al., 2014), it sheds little light into the socio-technical context of designing and adopting PHI tools within community organizations. To address this gap in research, human–computer interaction (HCI) and computer-supported cooperative work researchers have begun to conduct socio-technical research on community-based health technologies. Much of this work has focused on schools (Berkovsky et al., 2012; Lee et al., 2015; Macvean and Robertson, 2013), including evaluations of wearable and social computing tools that help students track progress toward health goals and engage with one another through social support, observational learning, and friendly competition (Miller and Mynatt, 2014; Poole et al., 2011). Outside of the school context, Maitland et al. (2009), explored design directions for nutrition promotion in public housing complexes. This research highlighted the criticality of understanding levels of trust among neighbors, and the implications for health technologies that enable social interactions within public housing settings. Also within the domain of social computing, *Community Mosaic* was a public interactive display system that helped residents send messages advocating healthy eating to others in their community (Parker et al., 2012; Parker and Grinter, 2014; Parker, 2014). An evaluation of this system showed that the consistent and public visibility of messages in the display empowered users to collectively advocate for change, counteracting the also consistent and visible draws to unhealthy eating in their neighborhoods.

Although CBOs offer valuable settings for intervention delivery, researchers have also noted constraints for those providing and utilizing the intervention, including financial, time, and transportation costs (Atkinson et al., 2010; King et al., 2013; Unertl et al., 2015). Such challenges can seem to counter the vision that ICTs will support lower-cost, accessible, and scalable health promotion solutions that reach a broader spectrum of people than has been possible with traditional approaches (King et al., 2013). An open question, then, is how can PHI systems be designed to minimize barriers to engagement while capitalizing on the benefits afforded by CBOs?

3. CASE STUDY

To answer this question and explore particular opportunities for family wellness promotion, we embarked upon a project that examines how technology can encourage physical activity in low-SES families. We partnered with Healthy Kids Healthy Futures (HKHF), an intergenerational obesity prevention initiative that promotes physical activity and healthy eating in caregivers (e.g., parents and grandparents) and children (Agrawal et al., 2012; Castaneda-Sceppa et al., 2014; Hoffman et al., 2012). One component of HKHF is Family Gym, a program that promotes physical activity in underserved families (e.g., low-income and racial and ethnic minority) with young children (3–8 years old). This initiative is offered in urban neighborhoods where play and recreational space are not easily accessible or safe.

Family Gym is offered year-round in three cycles of 8–10 sessions each during spring, summer, and fall. Sessions are held on Saturday mornings at three community center sites. During Family Gym, play equipment is set up in a large, open space to facilitate individual, small, and large-group play. Play activities were selected from evidence-based curricula to allow young children to participate and to engage caregivers as role models for their children (Fox et al., 2010; McKenzie et al., 1997; Sallis et al., 1997). Examples of activities include completing an obstacle course, jumping rope, playing limbo, soccer, basketball, and instructor-led Zumba and yoga sessions.

Family Gym attendance averages 56 children and adults per week. Program attendees reflect the demographics of Boston residents bearing a greater burden of chronic diseases such as obesity: the majority of participants come from neighborhoods with high poverty rates; 24% of participants are Latino and 43% are African American.

3.1 Formative Research

The authors began their collaboration in 2013 to explore how technology might expand the reach of the Family Gym program. While families are served by this program once per week, we saw an opportunity for technology to provide more ongoing encouragement for families to engage in physical activity inside and outside of Family Gym. We first conducted a formative study with caregivers attending Family Gym to understand their existing values around and attitudes toward physical activity, as well as families' current physical activity levels.

We conducted four focus group sessions with 13 caregivers (nine females, four males) whose children attended Family Gym. The median age of the caregivers was 36. We used a semistructured group interview guide to probe a variety of topics, including caregiver perspectives on child physical activity and being physically active with their child. The focus groups were recorded and transcribed, and we conducted an inductive, thematic analysis inspired by grounded theory (Corbin and Strauss, 2014). Two researchers independently coded the transcripts and met regularly to discuss and reconcile open codes. The researchers then iteratively clustered codes into higher-level themes and identified relationships between themes through axial coding.

We administered a follow-up survey to a subset of participants (n = 10), using validated instruments to gain more insight into a variety of physical activity attitudes and behaviors within families, such as: caregivers' physical activity stage of change (i.e., their position on a continuum from not thinking about increasing their physical activity to maintaining an increased level of activity; Steptoe et al., 2001), caregiver enjoyment of physical activity, their child's activity level, and the support caregivers provide for child activity (McMinn et al., 2009). Open-ended questions further probed participants' experiences at Family Gym, including aspects that they like and dislike. Descriptive statistics were computed for these data.

This research was approved by Northeastern University's Institutional Review Board. Participants received a $10 gift card for participating in the focus group and an additional $10 gift card for completing the follow-up survey.

3.2 Spaceship Launch

In our formative work, we found a mismatch between caregivers perception of their physical activity levels and their BMI—many self-reported being very active, yet most were overweight or obese (Saksono et al., 2015). These findings led us to design a system that (1) helps families understand the relationship between physical activity *intensity* and caloric burn and (2) encourages families to engage in increased moderate and vigorous physical activity (Saksono et al., 2015). The resulting system is a digital exergame called *Spaceship Launch*. Informed by social cognitive theory (Bandura, 2004), Spaceship Launch was designed to encourage caregivers and children to work together to be more physically active at a moderate or vigorous level. Spaceship Launch has three components: a physical activity data Dashboard, a Trivia Mini-Game, and the Launch Game (Figs. 4.1 and 4.2). The *Dashboard*

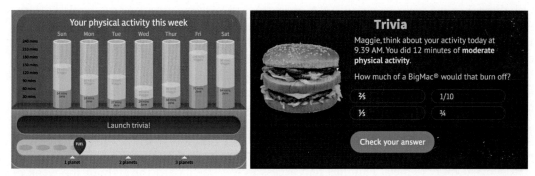

FIGURE 4.1 Spaceship Launch Dashboard (left) and Trivia Mini-Game (right). The Dashboard displays the time the caregiver and child spent in moderate and vigorous activity each day. Also displayed is the family's progress toward earning fuel to launch the child's spaceship to planets. The Trivia Mini-Game asks the child or caregiver to reflect on specific bouts of activity and consider how much food that activity would burn off.

FIGURE 4.2 Spaceship Launch Game. The child selects which planet he or she wants to travel to (left). Once the spaceship arrives at the planet, the child is sent on missions that involve being physically active (right). During Family Gym, this feature encourages children to be active and engage in the gym programming.

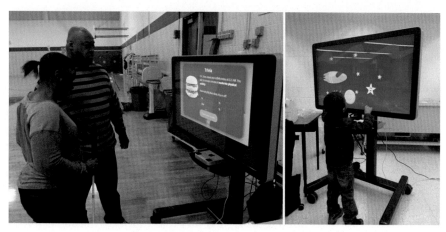

FIGURE 4.3   Adults and a child using Spaceship Launch on the interactive display installed at Family Gym.

visualizes the time caregivers and children spend in moderate and vigorous physical activity levels throughout the week. This information is visualized based upon data collected from Fitbit activity monitors worn by caregivers and children. Additionally, participants had periodic opportunities to play a *Trivia Mini-Game* within the Dashboard. This game asked them to recall specific bouts of activity during their day, and to guess how much food that activity would burn. This game attempts to scaffold increased literacy regarding the relationship between diet and physical activity.

The *Launch Game* provides families with digital rewards when they reach physical activity goals: time spent in moderate and vigorous activity by both the caregiver and child equates to fuel points in the game. As more fuel points are acquired, the child can launch his or her spaceship to more planets. Families had access to the Dashboard throughout the week on their personal devices (e.g., a home computer or a smartphone); the Launch game was accessible each Saturday at Family Gym on a large interactive touchscreen monitor and via participants' personal devices (Fig. 4.3).

3.3 Evaluation Study

We conducted a 3-week pilot study to evaluate user experience with Spaceship Launch. Twenty-nine people from 13 families participated: 15 caregivers and 14 children. Most caregivers (n=11) and children (n=10) were female. The median age of the caregivers was 39 and the median age of the children was 8. This research was approved by Northeastern University's Institutional Review Board. Participants received a gift card for $20–$40 (the amount was determined based upon the number of interviews completed).

At the beginning of the study, participants completed a validated survey to assess their intention to be physically active in the next month (Courneya and McAuley, 1993), physical activity stage of change (Meriwether et al., 2006), and support for child physical activity (McMinn et al., 2009). Each caregiver was given a Fitbit Flex and each child was given a Fitbit Zip to wear. Families wore the Fitbits for 1 week to collect baseline data and help families become accustomed to wearing the devices. At the end of this baseline period, participants were given access to Spaceship Launch for 3 weeks. We conducted semistructured interviews with eight caregivers to examine their experience with the game and how it affected their PA intention and modeling behaviors. We also conducted participatory design workshops with five caregivers and four children to probe how the game could better encourage physical activity.

The interviews and design workshops were audio recorded and transcribed. We conducted an inductive, thematic analysis of this data inspired by grounded theory (Corbin and Strauss, 2014): we iteratively coded transcripts and clustered codes to arrive at higher-level themes, using axial coding to identify relationships between themes. Descriptive statistics were computed for the survey data.

4. FINDINGS

We next discuss findings from our formative study, how these findings led to the design of Spaceship Launch, and results from our pilot evaluation of Spaceship Launch. Findings from this project have been described elsewhere (Saksono et al., 2015). In this chapter, we review previously reported findings and introduce new results to highlight the importance of a socio-ecological approach to community-based health technology design. Such an approach involves studying the intersection of people and their social, organizational, and physical environments—and how these relationships inhibit or support health behaviors and attitudes (Stokols, 1996). We extend the typical application of a socio-ecological framework in the health sciences to also include an examination of how health technologies can be designed to mediate these relationships. We discuss implications for family-based health technologies that arose as we evaluated intersecting needs, attitudes, values, and other phenomena between families and the CBO we worked with.

We begin with a discussion of *intra-family* values and practices and how these *aligned and diverged from Family Gym values*. We then describe how the *physical and operational facets of Family Gym* provided opportunities to address the technical problems that our families faced as well as the challenges that arose from working within this dynamic context. When reporting quotes from our participants, we use the prefixes "F" and "E" in participant IDs to refer to caregivers in our formative and evaluation studies, respectively. In addition, we use the term *caregivers* to describe the adult participants in our studies collectively: while most participants were parents, our sample also included a grandparent and an uncle.

4.1 Family and Organizational Values: Concordance and Dissonance

When designing systems for lay users within any organizational setting, it is important to understand how service users' values intersect with the organization's values. Identifying concordance and dissonance in these values helps designers to create systems that meet the varied stakeholder needs and desires. In our project, two salient stakeholder groups were the families who attended Family Gym and the staff and researchers who administered the program. Through our formative work, we identified the nuanced ways in which families' health attitudes and behaviors aligned with and diverged from the values of the Family Gym program.

In many ways, caregivers shared the values espoused and promoted by Family Gym. For example, in line with the program's emphasis on physical activity as a means of preventing obesity, participants expressed awareness of the health risks associated with obesity and valued physical activity as a method of prevention. The program also seeks to encourage physical activity enjoyment in families. In line with this value, caregivers reported feeling a sense of joy and satisfaction when they saw their child being active, and a sense of connectedness when they are active together. P4 expressed that she can be carefree, without inhibition, when active with her child:

> F4: I also feel silly [when exercising with my child] because we're having a good time. We're real goofy together.

Beyond enjoyment, our findings helped surface additional, complimentary emotional experiences. For example, P2, who is visually impaired, feels accepted when her daughter wants to be active with her:

> F2: [I feel] proud because my daughter is not ashamed of me. I can't see... and it doesn't bother her to have me there.

Other caregivers saw physical activity as an opportunity to transfer values and skills to prepare their kids for adulthood:

> F3: I feel like I'm teaching him something. I'm learning what he enjoys to do, what he likes and doesn't like. Also, I feel like I'm coaching him, coaching him the right way, the proper way of doing things.

These findings helped characterize affective properties of families' wellness pursuits, and the importance of accounting for such factors when creating family-based health technologies. Indeed, in our evaluation study, a significant finding was that some families valued Spaceship Launch because of the opportunities it gives families to provide and receive words of encouragement and affirmation, and to bond (Saksono et al., 2015).

While encouraging their children to be physically active and connecting emotionally with their children were strong values for families, there was less alignment with other Family Gym values. For example, Family Gym encourages caregiver behavioral modeling—a form of social support whereby show caregivers show their children that they are active themselves. Caregiver modeling supports child physical activity by building up positive values in children. Modeling is encouraged in Family Gym through the provision of activities that caregivers and children can do together, and via informational materials (e.g., announcements and handouts). However, we found that *live modeling* of physical activity (i.e., being active in front of children) was the form of support least provided by caregivers in our formative study, and it was tied for least-provided form of support in our evaluation study (Tables 4.1 and 4.2). In our evaluation study, we also assessed *verbal modeling* (in which caregivers discuss their prior physical activity with kids) and found that this was also done infrequently (Table 4.2). A prior study similarly found that caregivers were rarely physically active during Family Gym sessions (e.g., they were instead talking with other caregivers or standing; Castaneda-Sceppa et al., 2014). These findings suggest that for many families, physical activity modeling was not a prevalent behavior inside and outside of Family Gym.

TABLE 4.1   Caregiver Support for Child Physical Activity (Formative Study)

Frequency of Caregiver Support: Formative Study	Median	IQR	Min	Max
Tell kids the value of physical activity	5	0	4	5
Encouraged kids to be physically active	5	0	4	5
Provided transportation to activities	5	0.75	3	5
Watched kids be physically active	4	1.75	3	5
Participate in physical activity with kids (live modeling)	3.7	1.75	2	5

This table describes the frequency with which caregivers engaged in various forms of social support. Caregivers indicated how frequently they provided each form of support on a 5-point scale (1: never, 5: daily). For each category of support, we list the median, interquartile range (IQR), minimum and maximum response values.

TABLE 4.2   Caregiver Support for Child Physical Activity (Evaluation Study)

Frequency of Caregiver Support: Evaluation Study	Median	IQR	Min	Max
Tell kids the value of physical activity	4	1	1	5
Encouraged kids to be physically active	3	1	1	5
Provided transportation to activities	3	1	1	5
Tell kids about caregiver activity (*verbal modeling*)	3	1	1	5
Watched kids be physically active	3	0.5	1	5
Participate in physical activity with kids (*live modeling*)	3	0.5	1	5

This table describes the frequency with which caregivers engaged in various forms of social support. Caregivers indicated how frequently they provided each form of support on a 5-point scale (1: never, 5: daily). For each category of support, we list the median, interquartile range (IQR), minimum and maximum response values.

While modeling was low, many caregivers were clearly interested in being active with their children: when asked what they liked about Family Gym in the formative survey, several caregivers discussed valuing the opportunity to be active with their children and model positive behaviors. There are many potential reasons for the limited caregiver modeling that we observed. Brownson et al. (2001) found that the top physical activity barriers for lower-income adults are the perceptions that one has had enough activity at work, being too tired, and not having time. Following the recommendations proposed by Thompson et al. (2010) to develop health interventions that accommodate the complex needs and demands of today's families, we see potential for health technologies that help caregivers reflect on their prior activity with their children (even if they are not active at the same place and time) *and use that reflection process as a modeling opportunity.*

Spaceship Launch was our initial exploration of this design space. Our evaluation study showed that the system did support caregivers and children in collectively assessing their physical activity levels. Family members compared their activity levels to one another and caregivers used the reflective opportunity to provide positive feedback and instill positive values in their children. For example, caregivers were able to discuss their own activity levels with their children, pointing out when they had accomplished a healthy amount of activity for the day. E9 discussed viewing the Dashboard with her kids this way:

> E9: They did not know that I was moving that much. Well, neither did I. I didn't know really. I know I move but... It's different when you see it in minutes... So then they get to see, and they're like, "Wow! You move that much?" I say, "Yeah." So every night they kind of wait for me so we go into trivia, see how much I've done for the day. And they kind of have their own comments to say, "Oh, you didn't do much today." Or, "You did much today." ... They're like, "Oh, no wonder now you're losing weight." Even though I didn't lose pound wise... But I have a lot of energy now too, so. They're happy. And I keep going. They say, "Keep going mommy."

This quote exemplifies the benefit that computer-mediated displays can provide families, making visible caregivers' activities to children and enabling modeling opportunities even when caregivers and kids are not together. While many participants discussed the benefit of a tool like Spaceship Launch for encouraging child wellness, E9's comments highlight how such tools can also enable meaningful praise and encouragement from the child to the caregiver. E9 went onto further discuss the emotional benefits of sharing her activity levels with her children:

> E9: I think I'm proud for them that they recognize the difference...of like how, my body difference for example... Getting in shape better. They notice that I'm active... And it's kind of like nice that they notice.

Some families also described the benefit of letting children know when they had surpassed their caregivers' activity levels, indicating that this is a valuable opportunity to provide praise to the child:

> E1: I'll go, "Look, you won...you did more activity... You get to the planet more quicker than I did, because you did XYZ." And he'll go, "Oh, you need to exercise more mommy"... And [by observing the PA collectively, we] may encourage each other: "We've got to do more". Or just like, "Go, keep doing the same thing we're doing right now."

Our findings suggest that exergames and other interactive systems may be acceptable and viable ways to support the caregiver modeling behaviors encouraged by the Family Gym program. Such tools can complement the in-person resources provided by the program to accommodate the reality that families are often apart during the day, which limits opportunities to model positive behaviors in-person. At the same time, E4 discussed how time still acted as a barrier for her to review the Spaceship Launch Dashboard with her child. E4 would typically review the Dashboard when she is at work or when her child is asleep and noted that there is very little time when they are together during the day and at a computer to the view the Dashboard. These comments suggest the importance of further research to examine how health technologies can be designed to fit more seamlessly within the varied routines of families.

In summary, examining the alignment and divergence of the Family Gym and caregiver values helped us establish important design directions for our Spaceship Launch software. In taking this approach, we were able to create a tool that works in concert with the existing Family Gym infrastructure. (Later in this chapter we discuss the benefits and challenges of this approach.) With Spaceship Launch, our goals were to evaluate how a tool that encourages the development of physical activity values and behaviors promoted by the Family Gym program can be done in a way that families feel is enjoyable and engaging. In our evaluation study, caregivers reported an increased awareness and discussion of activity levels with their children and ways in which the game motivated them to increase their physical activity. Based upon these findings, we see family-based health technologies such as Spaceship Launch as promising platforms for supporting behavioral modeling and change in families.

4.2 Physical Affordances

In contrast to formal health care settings that people may only visit when sick, neighborhood institutions such as community centers, schools, churches, and libraries offer programming and resources that draw residents on a more regular basis. For example, Family Gym was offered within a community center that residents attended each week. This regularity enabled us to create an application that leveraged a situated display approach. Working within the shared, open space of the community center was beneficial because it allowed us to address caregivers' technology literacy challenges and create a tool that was physically embedded within a real-world physical activity program.

First, technology literacy was a major challenge, as seven of the thirteen families had problems setting up the software driver that enables the Fitbits to wirelessly sync their data with their home computer. At the start of the study, we offered to assist each family with this home installation, but only two accepted the offer. Another frequently occurring problem was that families would accidentally disconnect their Fitbits from the game by setting up their own private Fitbit account. While, for years, ubiquitous computing researchers have grappled with installation and other technical challenges in domestic computing research, this problem becomes exacerbated when working with low-SES families, where digital literacy may be even lower (Ginossar and Nelson, 2010; Greenhow et al., 2009; Stanley, 2003).

Technical challenges such as those faced by our participants can act as a barrier to engagement with health technologies. To counteract these challenges, one promising direction is to provide access to health technologies in public community settings. This approach can increase the accessibility of PHI interventions for families with limited digital literacy or technology access at home. Spaceship Launch was accessible both on caregivers' personal devices and at Family Gym. We found that more families accessed the software at the gym than via personal devices such as a computers and smartphones, suggesting the merit of providing multi-modal access to health technologies for this demographic. This recommendation is supported by research demonstrating that children in low-income families spend less time with a home computer than children in higher-income households (Roberts, 2000), and low-income households are more likely than higher-income households to access the Internet outside of the home (Zickuhr, 2013). Furthermore, low computer literacy acts as a barrier to Internet use, even more so than the cost of Internet access and a lack of access (Zickuhr, 2013). Researchers must carefully consider the sustainability of health technologies that are anchored within CBOs, creating solutions that can be easily maintained without adding excessive maintenance overhead for program staff.

A second benefit of integrating Spaceship Launch into a physical community space was that we were able to create an engaging software experience that was tightly paired with a real-world physical activity promotion setting. Indeed, in our observations of Spaceship

Launch, we found that children were frequently crowded around the display, eager to interact with the tool. Yet our design approach (namely that the game supports very brief interactions and then requires children to be physically active for an extended period of time before they can play again) was such that users were driven back into the real-world physical activity programming. This approach helped us leverage the power of digital media for engaging and incentivizing healthy behaviors, while still encouraging users to take advantage of real-world resources for physical activity.

4.3 Operational Factors

There were several facets of the Family Gym program that impacted the design and delivery of Spaceship Launch, as well as our evaluation of the tool. First, as the Family Gym program is staffed in part by University students, the program is offered in cycles that align with the school year. Sessions of 8–10 weeks are provided during the fall, spring, and summer semesters. Given that we were designing a tool that would be used in the context of this program, its periodicity constrained our data collection and evaluation study design. For example, 63% of families attend only one cycle of Family Gym, which meant that carrying our study over multiple Family Gym cycles could lead to significant attrition in the study. Furthermore, the gap between gym cycles could further impact participant engagement with the system. Community-based programming is often cyclic or subject to starts and stops for a variety of reasons. One direction for future research is to create software that deliberately accounts for these gaps, providing users with access to content and features in the interim (e.g., via personal devices). Another direction would be to design tools that more explicitly attempt to sustain engagement with the organization, reducing attrition through engaging system content and supports.

Many of the issues we encountered reflect fundamental concerns in any community-based research project, namely understanding and accounting for organizational constraints and concerns. However, the issues we faced raise particular challenges within the context of health technology design. The dynamic nature of community-based programming (e.g., in terms of its periodicity, duration, and financial constraints) means that technologies must be carefully designed to be flexible and adaptive to such changes. When creating systems to encourage health behavior change (a process that happens over time), it may be critical to provide sustained supports to users that endure despite fluctuations in programming.

5. DISCUSSION

This case study characterizes the socio-ecological context of designing health technologies that work in conjunction with CBOs—the ways that community residents and the organizational environment intersect to produce opportunities and challenges for design. A cross-cutting theme in our findings is that to design effective health technologies, it is critically important to understand the values, needs, practices, and attitudes of families and how these factors intersect with the values, priorities, affordances, and constraints of CBOs. We conclude with directions for future PHI research in community settings.

5.1 The Value of a Community-Based Approach for Addressing Issues of Health Equity

Low-SES populations face added financial, informational, social, and environmental barriers to wellness. Designing technologies within a CBO context can help residents overcome these barriers in several ways. First, to address concerns regarding the costs of new technological interventions within resource poor environments, a promising approach is to centralize spending on technology that is housed at the CBO. One model would be to provide a standard software interface on more widespread personal devices (e.g., phones) and supplementary functionality that leverages more sophisticated or costly hardware at the CBO. This approach not only reduces the costs incurred by CBO members but also the technological complexity that they encounter. With Spaceship Launch, we offered an engaging digital media experience with the game via a large interactive display installed in the community center. Rather than require families have individual ownership of expensive displays, we were able to provide a shared resource to minimize costs.

While our approach did introduce the use of wearable activity monitors, which bring added cost, our vision is that in the future, such costs will be mitigated by approaches such as utilizing activity sensors in more widely owned smartphone devices. Innovative initiatives may further enable community-based wellness technologies; for example, RecycleHealth is a nonprofit organization to which consumers donate wearable devices that they no longer use. The trackers then get distributed to organizations serving vulnerable populations (e.g., older adults and low-income families; Ducharme, 2016). Creative programs such as these, coupled with a centralized model that alleviates financial and technical burden, present a promising approach to delivering community-based health solutions.

Second, many resource-poor cities and towns have existing CBOs that are providing varied, valued, and needed services. For example, affordable produce programs, nutrition classes, walking programs, and financial literacy courses are all examples of resources that health centers, community centers, libraries, and grassroots organizations across the country offer to low-SES neighborhoods. These resources address the range of physical, social, educational, and financial supports that families need to achieve a state of holistic wellbeing. However, services are often provided in silos and residents may not be aware of or take advantage of these opportunities—leaving these resources underutilized. Technological innovations can help better connect residents to the health and wellness services already offered by CBOs, augmenting, magnifying, and extending the reach of these institutions.

For example, Spaceship Launch worked in tandem with the Family Gym program, providing a bridge between Saturday gym sessions through the physical activity encouragement and reflection opportunities offered in the application. Future research might explore how technology can facilitate access to programs that residents are unable to attend in person (e.g., remote presence systems that support distributed participation in educational workshops) or that help residents build upon experiences in CBO events (e.g., tools that help residents document and share how they iterate upon recipes learned in nutritious cooking workshops). Indeed, this type of strengths-based approach (in which community assets are identified and leveraged) is needed to balance a focus on the challenges that exist in local neighborhoods (Israel et al., 1998).

5.2 Untapped Potential: Leveraging CBPR Models of Neighborhood Health Promotion

Within public health, CBPR is an approach that addresses "social, structural, and physical environmental inequities through active involvement of community members, organizational representatives, and researchers in all aspects of the research process" (Israel et al., 1998). CBPR has taken many forms, each of which can serve as a guide for the empirical study and design of community-based health technologies. For example, there is a large literature on faith-based interventions, in which health programs are developed with and delivered in the context of churches and other faith-centered organizations (Campbell et al., 2007; DeHaven et al., 2004). The benefits of a faith-based approach include opportunities to build upon the health and wellness initiatives that are a part of many religious institutions' missions. Furthermore, these entities provide an opportunity to recruit a stable set of participants longitudinally, given that parishioners often attend a church frequently over many years (Campbell et al., 2007). In the United States, these entities provide access to a broad swath of the population (most Americans attend church or another religious institution), and offer unique opportunities to address disparities within populations such as African Americans, where participation in faith communities is particularly strong (Campbell et al., 2007).

Yet, even given these and many other affordances of faith-based organizations they have remained an unexplored domain for HCI health researchers. Outside of HCI, some researchers have begun to explore this space. For example, the *Guide to Health* intervention includes online nutrition and physical activity education and goal-setting tools (Winett et al., 2007). The weekly rhythm of church services presents opportunities for engagement: reminders to use the intervention resources are included in church services and bulletins, and churchwide behavior change goals provide socially relevant motivation. Still, faith-based health technologies are a nascent area of research, to which the field of HCI has much to contribute. In particular, research is needed to explicate how technologies designed for faith-based contexts can enhance user experience and long-term adoption. Such research would integrate the currently disparate research agendas exploring technology's role in spirituality (Blythe and Buie 2014; Gaver et al., 2010; Uriu and Odom 2016; Wyche et al., 2009) and wellness within HCI, and allow the pursuit of new questions about the intersection of faith, health, and technology use.

While exploring the design space for faith-based organizations offers significant opportunities, it also presents important challenges. For example, intervention designers must be careful to respect the values espoused by churches, not trivializing them and being careful to accurately reflect them in intervention design (Campbell et al., 2007). Other challenges arise when the ideals, mission, or tenets of a faith tradition conflict with those of the researchers (Campbell et al., 2007), adding further complexity to the participatory, user-centered design process that is standard practice within HCI. Value sensitive design is an approach that HCI researchers have used to address a range of domains and could prove a valuable methodological orientation for health technology research in faith-based contexts (Friedman et al., 2013).

Faith-based interventions represent just one of several approaches within CBPR. A vibrant community-based PHI agenda will explore these and other opportunities for design, such as technologies that support community organizing (whereby residents are empowered to collectively implement strategies addressing local wellness barriers; Minkler and Wallerstein,

2005). Developing and delivering interventions within CBOs is a promising way of increasing community empowerment and ownership of health interventions, which can lead to higher rates of participation and long-term sustainability (Campbell et al., 2007). These benefits are particularly exciting for health technology researchers, offering a promising opportunity to overcome the high and consistent rates of attrition documented for ICTs that seek to encourage healthy behaviors (Jimison et al., 2008; Rodgers et al., 2016).

6. CONCLUSION

In this chapter, we described the design and evaluation of an exergame for families. This case study characterized benefits and challenges of creating health technologies within the community-based organizational context. While health technology research in HCI and related disciplines has focused primarily on everyday computing solutions (used apart from any particular institution) or systems within clinical settings, there is a compelling opportunity to pursue a research agenda focused on CBOs. CBOs offer significant opportunities to sustain user engagement and address persistent health disparities by reaching vulnerable populations underserved by traditional health care institutions. More work is needed to develop models of innovation and dissemination in this domain and to evaluate the adoption and impact of community-based health systems longitudinally. Finally, comparative effectiveness studies will be critical to demonstrate how various approaches to design support user engagement and healthy behaviors within and across populations and in different types of CBOs.

Acknowledgments

The authors thank the Health Kids, Healthy Futures, and Family Gym staff for their support of this project.

References

Agrawal, T., Hoffman, J.A., Ahl, M., Bhaumik, U., Healey, C., Carter, S., Dickerson, D., Nethersole, S., Griffin, D., Castaneda-Sceppa, C., 2012. Collaborating for impact: a multilevel early childhood obesity prevention initiative. Family and Community Health 35 (3), 192–202.

Atkinson, N.L., Desmond, S.M., Saperstein, S.L., Billing, A.S., Gold, R.S., Tournas-Hardt, A., 2010. Assets, challenges, and the potential of technology for nutrition education in rural communities. Journal of Nutrition Education and Behavior 42 (6), 410–416.

Bandura, A., 2004. Health promotion by social cognitive means. Health Education and Behavior 31 (2), 143–164.

Berkovsky, S., Freyne, J., Coombe, Mac, 2012. Physical activity motivating games: be active and get your own reward. ACM Transactions on Computer-human Interaction 19 (4), 32.

Bers, M.U., Cantrell, K., 2012. Virtual worlds for children with medical conditions: experiences for promoting positive youth development. In: Smedberg, Å. (Ed.), E-health Communities and Online Self-help Groups: Applications and Usage. IGI Global, Hershey, PA, pp. 1–23.

Bers, M.U., Gonzalez-Heydrich, J., DeMaso, D.R., 2001. Identity construction environments: supporting a virtual therapeutic community of pediatric patients undergoing dialysis. In: Proceedings of the SIGCHI Conference on Human Factors in Computing Systems, pp. 380–387.

Blythe, M., Buie, E., 2014. Chatbots of the gods: imaginary abstracts for techno-spirituality research. In: Proceedings of the 8th Nordic Conference on Human-computer Interaction: Fun, Fast, Foundational, pp. 227–236.

Brownson, R.C., Baker, E.A., Housemann, R.A., Brennan, L.K., Bacak, S.J., 2001. Environmental and policy determinants of physical activity in the United States. American Journal of Public Health 91 (12), 1995–2003.

Campbell, Kramish, M., Allicock Hudson, M., Resnicow, K., Blakeney, N., Paxton, A., Baskin, M., 2007. Church-based health promotion interventions: evidence and lessons learned. Annual Review of Public Health 28, 213–234.

Castaneda-Sceppa, C., Hoffman, J.A., Thomas, J., DuBois, M., Agrawal, T., Griffin, D., Bhaumik, U., Locke Healey, C., Dickerson, D., Nethersole, S., 2014. Family gym: a model to promote physical activity for families with young children. Journal of Healthcare for the Poor and Underserved 25 (3), 1101–1107.

Centers for Disease Control and Prevention, 2011. Principles of Community Engagement.

Chen, F.X., King, A.C., Hekler, E.B., 2014. Healthifying exergames: improving health outcomes through intentional priming. In: Proceedings of the SIGCHI Conference on Human Factors in Computing Systems, pp. 1855–1864.

Consolvo, S., Klasnja, P., McDonald, D.W., Avrahami, D., Froehlich, J., LeGrand, L., Libby, R., Mosher, K., Landay, J.A., 2008. Flowers or a robot army?: encouraging awareness & activity with personal, mobile displays. In: Proceedings of the 10th International Conference on Ubiquitous Computing, pp. 54–63.

Corbin, J., Strauss, A., 2014. Basics of Qualitative Research: Techniques and Procedures for Developing Grounded Theory. Sage publications.

Cordeiro, F., Bales, E., Cherry, E., Fogarty, J., 2015. Rethinking the mobile food journal: exploring opportunities for lightweight photo-based capture. In: Proceedings of the 33rd Annual ACM Conference on Human Factors in Computing Systems.

Courneya, K.S., McAuley, E., 1993. Predicting physical activity from intention: conceptual and methodological issues. Journal of Sport and Exercise Psychology 15 (1), 50.

DeHaven, M.J., Hunter, I.B., Wilder, L., Walton, J.W., Berry, J., 2004. Health programs in faith-based organizations: are they effective? American Journal of Public Health 94 (6), 1030–1036.

Ducharme, J., 2016. This Group Gives Unused Fitness Trackers Second Lives. Boston Magazine.

El-Nasr, M.S., Durga, S., Shiyko, M., Sceppa, C., 2015. Unpacking Adherence and Engagement in Pervasive Health Games. Foundations of Digital Games.

Epstein, D., Cordeiro, F., Bales, E., Fogarty, J., Munson, S., 2014. Taming data complexity in lifelogs: exploring visual cuts of personal informatics data. In: Proceedings of the 2014 Conference on Designing Interactive Systems, pp. 667–676.

Fox, M.K., Hallgren, K., Boller, K., Turner, A., Cabili, C., Condon, E., Del Grosso, P., Eden, D., Finkelstein, D., Kennen, B., 2010. Efforts to Meet Childrens Physical Activity and Nutritional Needs Findings from the I Am Moving I Am Learning Implementation Evaluation. Mathematica Policy Research.

Friedman, B., Kahn Jr., P.H., Borning, A., Huldtgren, A., 2013. Value sensitive design and information systems. In: Early Engagement and New Technologies: Opening up the Laboratory. Springer, pp. 55–95.

Gaver, W., Blythe, M., Boucher, A., Jarvis, N., Bowers, J., Wright, P., 2010. The prayer companion: openness and specificity, materiality and spirituality. In: Proceedings of the SIGCHI Conference on Human Factors in Computing Systems, pp. 2055–2064.

Ginossar, T., Nelson, S., 2010. Reducing the health and digital divides: a model for using community-based participatory research approach to e-health interventions in low-income hispanic communities. Journal of Computer-Mediated Communication 15 (4), 530–551.

Glanz, K., Rimer, B.K., Viswanath, K. (Eds.), 2008. Health Behavior and Health Education. Jossey-Bass, San Francisco, CA.

Greenhow, C., Walker, J.D., Kim, S., 2009. Millennial learners and net-savvy teens? Examining Internet use among low-income students. Journal of Computing in Teacher Education 26 (2), 63–68.

Hoffman, J.A., Agrawal, T., Wirth, C., Watts, C., Adeduntan, G., Myles, L., Castaneda-Sceppa, C., 2012. Farm to family: increasing access to affordable fruits and vegetables among urban head start families. Journal of Hunger and Environmental Nutrition 7 (2–3), 165–177.

Israel, B.A., Schulz, A.J., Parker, E.A., Becker, A.B., 1998. Review of community-based research: assessing partnership approaches to improve public health. Annual Review of Public Health 19 (1), 173–202.

Jimison, H., Gorman, P., Woods, S., Nygren, P., Walker, M., Norris, S., Hersh, W., 2008. Barriers and Drivers of Health Information Technology Use for the Elderly, Chronically Ill, and Underserved. Agency for Healthcare Research and Quality, Rockville, MD.

Kay, M., Choe, E.K., Shepherd, J., Greenstein, B., Watson, N., Consolvo, S., Kientz, J.A., 2012. Lullaby: a capture & access system for understanding the sleep environment. In: Proceedings of the 2012 ACM Conference on Ubiquitous Computing, pp. 226–234.

King, A.C., Hekler, E.B., Grieco, L.A., Winter, S.J., Sheats, J.L., Buman, M.P., Banerjee, B., Robinson, T.N., Cirimele, J., 2013. Harnessing different motivational frames via mobile phones to promote daily physical activity and reduce sedentary behavior in aging adults. PLoS One 8 (4), e62613.

Lee, V.R., Drake, J.R., Cain, R., Thayne, J., 2015. Opportunistic uses of the traditional school day through student examination of Fitbit activity tracker data. In: Proceedings of the 14th International Conference on Interaction Design and Children, pp. 209–218.

Macvean, A., Robertson, J., 2013. Understanding exergame users' physical activity, motivation and behavior over time. In: Proceedings of the SIGCHI Conference on Human Factors in Computing Systems, pp. 1251–1260.

Maitland, J., Chalmers, M., Siek, K., 2009. Persuasion not required: improving our understanding of the sociotechnical context of dietary behavioural change. In: 3rd International Conference on Pervasive Computing Technologies for Healthcare, 2009, pp. 1–8.

Matthews, M., Doherty, G., 2011. In the mood: engaging teenagers in psychotherapy using mobile phones. In: Proceedings of the SIGCHI Conference on Human Factors in Computing Systems, pp. 2947–2956.

McKenzie, T.L., Sallis, J.F., Kolody, B., Nell Faucette, F., 1997. Long-term effects of a physical education curriculum and staff development program: SPARK. Research Quarterly for Exercise and Sport 68 (4), 280–291.

McMinn, A., Sluijs, E.M., Harvey, N., Cooper, C., Inskip, H., Godfrey, K., Griffin, S., 2009. Validation of a maternal questionnaire on correlates of physical activity in preschool children. International Journal of Behavioral Nutrition and Physical Activity 6 (1), 81.

Meriwether, R.A., McMahon, P.M., Islam, N., Steinmann, W.C., 2006. Physical activity assessment: validation of a clinical assessment tool. American Journal of Preventive Medicine 31 (6), 484–491.

Miller, A.D., Mynatt, E.D., 2014. StepStream: a school-based pervasive social fitness system for everyday adolescent health. In: Proceedings of the 32nd Annual ACM Conference on Human Factors in Computing Systems, pp. 2823–2832.

Miller, A.D., Mishra, S.R., Kendall, L., Haldar, S., Pollack, A.H., Pratt, W., 2016. Partners in care: design considerations for caregivers and patients during a hospital stay. In: Proceedings of the 19th ACM Conference on Computer-supported Cooperative Work & Social Computing, pp. 756–769.

Minkler, M., 2005. Community-based research partnerships: challenges and opportunities. Journal of Urban Health 82 (2), ii3–ii12.

Minkler, M., Wallerstein, N., 2005. Improving health through community organization. In: Community Organizing and Community Building for Health. Rutgers University Press, pp. 26–51.

Minkler, M., Wallerstein, N., 2008. Community-Based Participatory Research for Health: from Process to Outcomes. John Wiley & Sons, San Francisco.

Nollen, N.L., Hutcheson, T., Carlson, S., Rapoff, M., Goggin, K., Mayfield, C., Ellerbeck, E., 2013. Development and functionality of a handheld computer program to improve fruit and vegetable intake among low-income youth. Health Education Research 28 (2), 249–264.

Parker, A.G., Kantroo, V., Lee, H., Osornio, M., Sharma, M., Grinter, R.E., 2012. Health promotion as activism: building community capacity to affect social change. In: Proceedings of the SIGCHI Conference on Human Factors in Computing Systems, pp. 99–108.

Parker, A.G., Grinter, R.E., 2014. Collectivistic health promotion tools: accounting for the relationship between culture, food and nutrition. International Journal of Human-Computer Studies 72 (2), 185–206.

Parker, A.G., 2014. Reflection-through-performance: personal implications of documenting health behaviors for the collective. Personal and Ubiquitous Computing 18 (7), 1737–1752.

Poole, E.S., Miller, A.D., Xu, Y., Eiriksdottir, E., Catrambone, R., Mynatt, E.D., 2011. The place for ubiquitous computing in schools: lessons learned from a school-based intervention for youth physical activity. In: Proceedings of the 13th International Conference on Ubiquitous Computing, pp. 395–404.

Roberts, D.F., 2000. Kids and Media at the New Millennium. Diane Publishing.

Rodgers, R.F., Franko, D.L., Shiyko, M., Intille, S., Wilson, K., O'Carroll, D., Lovering, M., Matsumoto, A., Iannuccilli, A., Luk, S., 2016. Exploring healthy eating among ethnic minority students using mobile technology: feasibility and adherence. Health Informatics Journal 22 (3), 440–450.

Saksono, H., Ashwini, R., Geeta, K., Carmen, C.-S., Hoffman, J.A., Wirth, C., Parker, A.G., 2015. Spaceship launch: designing a collaborative exergame for families. In: Proceedings of the 18th ACM Conference on Computer Supported Cooperative Work & Social Computing, pp. 1776–1787.

Sallis, J.F., McKenzie, T.L., Alcaraz, J.E., Kolody, B., Faucette, N., Hovell, M.F., 1997. The effects of a 2-year physical education program (SPARK) on physical activity and fitness in elementary school students. Sports, Play and Active Recreation for Kids. American Journal of Public Health 87 (8), 1328–1334.

Skeels, M., Tan, D.S., 2010. Identifying opportunities for inpatient-centric technology. In: Proceedings of the 1st ACM International Health Informatics Symposium, pp. 580–589.

Smith, J.J., Morgan, P.J., Plotnikoff, R.C., Dally, K.A., Salmon, J., Okely, A.D., Finn, T.L., Lubans, D.R., 2014. Smartphone obesity prevention trial for adolescent boys in low-income communities: the ATLAS RCT. Pediatrics 134 (3), e723–e731.

Stanley, L.D., 2003. Beyond access: psychosocial barriers to computer literacy special issue: ICTs and community networking. The Information Society 19 (5), 407–416.

Steptoe, A., Kerry, S., Rink, E., Hilton, S., 2001. The impact of behavioral counseling on stage of change in fat intake, physical activity, and cigarette smoking in adults at increased risk of coronary heart disease. American Journal of Public Health 91 (2), 265.

Stokols, D., 1996. Translating social ecological theory into guidelines for health promotion. American Journal of Health Promotion 10 (4), 282–298.

Thompson, J.L., Jago, R., Brockman, R., Cartwright, K., Page, A.S., Fox, K.R., 2010. Physically active families–debunking the myth? A qualitative study of family participation in physical activity. Child: Care, Health and Development 36 (2), 265–274.

Unertl, K.M., Schaefbauer, C.L., Campbell, T.R., Senteio, C., Siek, K.A., Bakken, S., Veinot, T.C., 2015. Integrating community-based participatory research and informatics approaches to improve the engagement and health of underserved populations. Journal of the American Medical Informatics Association ocv094.

Uriu, D., Odom, W., 2016. Designing for domestic memorialization and remembrance: a field study of fenestra in Japan. In: Proceedings of the 2016 CHI Conference on Human Factors in Computing Systems, pp. 5945–5957.

Wilcox, L., Morris, D., Tan, D., Gatewood, J., 2010. Designing patient-centric information displays for hospitals. In: Proceedings of the SIGCHI Conference on Human Factors in Computing Systems, pp. 2123–2132.

Winett, R.A., Anderson, E.S., Wojcik, J.R., Winett, S.G., Bowden, T., 2007. Guide to health: nutrition and physical activity outcomes of a group-randomized trial of an Internet-based intervention in churches. Annals of Behavioral Medicine 33 (3), 251–261.

Wright, K., Newman Giger, J., Norris, K., Suro, Z., 2013. Impact of a nurse-directed, coordinated school health program to enhance physical activity behaviors and reduce body mass index among minority children: a parallel-group, randomized control trial. International Journal of Nursing Studies 50 (6), 727–737.

Wyche, S.P., Caine, K.E., Davison, B.K., Patel, S.N., Arteaga, M., Grinter, R.E., 2009. Sacred imagery in techno-spiritual design. In: Proceedings of the SIGCHI Conference on Human Factors in Computing Systems, pp. 55–58.

Zickuhr, K., 2013. Who's Not Online and Why. Pew Internet & American Life Project, p. 40.

Socio-technical Betwixtness: Design Rationales for Health Care IT

Claus Bossen
Aarhus University, Aarhus, Denmark

1. INTRODUCTION

In the broadest sense, socio-technical systems theory starts off with the assumption that the design and development of technology involves decisions upon how to distribute competences and functions between humans and technology. The design of technology does not only concern the internal mechanics or software, the layout of computer screens, and connecting various devices, but concerns also broader questions: should processes be automated, or overseen, and decided upon by humans? Which choice delivers the best security or the best quality? Should health examination results automatically be sent to patients' smartphones or personal health records, or are they better presented and explained by and in the presence of a health care professional? Should diabetic patients' glucose levels be monitored by an algorithm and an alert be send to the physician if the numbers are outside of set parameters? Or should the numbers be send to the patient, who decides whether to contact her physicians, since she herself knows best the context for and hence meaning of the numbers? Apart from minor updates and systems maintenance, designing and developing new technology is about redistributing functions and responsibilities in a socio-technical system (Baxter and Sommerville, 2011; Berg, 1999). For the same reason, designing and developing new technology goes beyond "supporting" work, because the socio-technical system as a whole most likely will change: implementing new technology in all likelihood means a new work practice and organizational change. Indeed, such change may be a purpose of a new system. Hence, new technology could be said to "constitute" rather than support work and organizations (Latour, 1992; Law, 1992). Design is inherently socio-technical and complicated.

Two complications will especially be at the center of attention in this chapter. One complication arises from the need to make choices between different goals, interests, and reasons for designing new technology, or in short a choice between different *design rationales*. Another complication arises from the fact that how we understand, describe, and represent the world, including the work practices, processes, and organizations into which new technologies are to be used, is always situated from a specific position and context. Our perception depends on how we generated data to produce such understandings, descriptions, and representations. Thus, in at least two ways the design and development of new technology such as health care IT, is in a position of betwixtness, having to make choices.

As for the first kind of betwixtness, a number of challenges immediately arise within the health care domain where there are multiple stakeholders and goals, and where the delivery of health care services even within a hospital involves the cooperation of a complex set of units, professions, and technologies: What is the goal of the envisioned socio-technical design? Who decides upon that, or facilitates the agreement on a consensus? Who will benefit from a redistribution of competences and functions? It is easy to agree on better quality for patients, and more efficiency from health care providers, but more difficult to agree on or foresee who will gain through, for example, less work, higher work life quality, or better pay, and who be set back and have to work harder and become less attractive as a work force. These questions become even more challenging whenever health care services require cooperation across organizational boundaries. Answering these questions implies making choices between one or more, and sometimes conflicting, *design rationales* based on implicit or explicit assumptions, arguments, and goals. In the case of health care IT, it has been argued that, for example, electronic health records serve multiple purposes: to distribute information across time and space to enhance clinical work; to make information upon diagnoses and treatment available to patients; and to generate information for reuse by management and boards of health (Greenhalgh et al., 2009). Similarly, telemedicine can be introduced with the purposes of enhance ongoing treatment and monitoring of patients; saving patients' time and costs for traveling to hospitals; reducing hospital costs by limiting the number of bed days; and accumulating information for medical research. Multiple rationales initially set designers and developers of IT systems in betwixt positions, since they have to choose whether to design for one or several purposes, and they may have to do stakeholder analysis to assess who has which stakes in the envisioned IT system, as well as their ability to further or hinder its success (see e.g., Eskerod et al., 2015). The choice of which goal(s) to aim for may of course be limited by whoever funds the development, and it is possible to "streamline" projects by focusing on one purpose or one stakeholder, rather than staying with the mess of multiple ones. However, the success of new technologies most often also depends on the appropriation of the system and reorganization of work practices by those using it. So even if top management funds and decides upon goals, technological design may have to take front-line staff and middle management into consideration. This latter point was one of the main insights coming out of the socio-technical studies of the Tavistock group in the 1950s (Trist, 1981), (see also, the classic text by Markus, 1983), and has been central to the field of participatory design that emerged in the late 1980s (Simonsen and Robertson, 2012). The point has also remained central to recent updates of the socio-technical approach (Baxter and Sommerville, 2011; Clegg, 2000; Fischer and Herrmann, 2011). Designing or redesigning socio-technical systems is an intervention into practices with multiple purposes and stakeholders.

As for the second kind of betwixtness, doing socio-technical design of new technologies necessarily builds upon an understanding about not only upon the future state to be strived for, but also upon an understanding about what the present state looks like. What do work practices and processes look like now, and what needs to be changed? The complication here is that answering the question of what is now, is situated in a specific position and context.

As Clegg (2000) states: " ... the introduction of a computer-based system that is used to book time against various projects may be seen as a means of achieving financial control by accountants, as a way of monitoring the work of its users by their line managers, and as a constraint on their autonomy and behaviour by the users themselves. Different people will interpret systems in different ways" (Clegg, 2000, p. 467). The description of a consultation at the doctor's office will differ whether one asks the patient, the secretary, the nurse, or the doctor. Whom it is relevant to ask depends of course upon what the new technology is meant to do.

The challenge of different interpretations is also a fundamental challenge for sociologies of work and how actions and tasks are perceived. This can be illustrated by discussing what it means to act. One view of actors and acting starts off focusing on the individual actors following set courses of action (Fig. 5.1, Top). However, looking closer acts come forward as involving a process of continuous action in which emergent circumstances and the interaction with other actors have to be monitored by the actor, who ongoingly has to adjust her actions to the contingencies arising: Instead of "an act", we get "interacting" requiring efforts of aligning, coordinating, and monitoring, and which is more dynamic and complex than in former the linear representation. Not only are there multiple goals, but actions interact and have to be adjusted to get at the intended goal(s) (Fig. 5.1, Bottom; for a fully developed argument and description on this position towards acting, see Strauss, 1993). Thus a linear and "rationalistic"—in the sense of simplifying and rationalizing—representation of action, can be contrasted with an interactional representation. This contrast will be of direct relevance to the first case to be presented below.

Representations of practices then should not be made too rashly and without detailed empirical knowledge if we want to avoid rationalistic representations. Even if we would like to rationalize an apparently "messy" practice, this should be done with caution, since streamlining work into linear, rational models entails the risk of ignoring or forgetting central features enabling those actions. Further, since no description of a phenomenon can capture all its aspects, but will highlight some and push others to the background, the act of representing requires making choices of what to make visible or not (Suchman, 1995).

The two kinds of betwixtness, choosing between design rationales and constructing understandings and representations, interact: Not only do different stakeholders pursue different

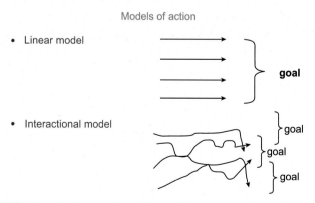

FIGURE 5.1 The linear, rationalistic, and the interactional models of action.

interests, but they may also have different perspectives and ways of representing work practices and organizational processes. It falls upon the socio-technical designer or developer of new technology to handle both multiple stakeholders and multiple representations of work. In caricature, this could, for example, involve choosing between doctors who act informed by medical, rational knowledge to make diagnoses and prescribe treatment for selected parts of patients' bodies; nurses informed by humanistic and hermeneutical approaches to humans in their care for the patient as a whole person; and medical secretaries dedicated to accuracy and completeness of information upon patients in the medical records.

In the following, this chapter will illustrate and discuss this twofold betwixtness by first presenting a brief historical outline of different design rationales in the design of technology and organizations, and second by analyzing two cases of health care IT development: The first case concerns the development of a basic model for electronic health records (EHRs), whereas the second case involves a logistic system for the coordination of hospital porter services. The discussion will then discuss the two cases and the betwixt position of design of IT in health care, and point at what may be learned.

2. A SHORT HISTORICAL BRIEF ON RATIONALES BEHIND THE DESIGN OF WORK, TECHNOLOGY, AND ORGANIZATIONS

The realization that design of work practices and organizations entails the combination of the social and the technical can be argued to go back at least to Taylor and the development of "scientific management" in the USA (Taylor, 1911). He posited his approach against a management style of "initiative and incentive" that used increased pay and positive encouragement by management to achieve efficiency. However, Taylor argued that such an approach of initiative and incentive left planning of work to the workers themselves, which led to inefficient organization, workers exhausting themselves before the work day was over, and suboptimal output. The famous and infamous approach of scientific management was instead based upon empirical studies of work based on which procedures, technologies, and organization were designed. Among other things, Taylor argued for the importance of regular breaks to make workers produce to maximum efficiency for a whole day: pushing workers too hard led to quick exhaustion and lower overall productivity, whereas regular breaks enabled workers to produce an overall larger output. Also, by making stop-watch observations of shoveling and varying the design of shovels, these could be designed with a size and length that meant that the load of coal for each thrust would not wear out the worker too quickly, but make him shovel with maximal daily output. Later, the design of technology was combined with studies of task movements so that the overall designed combination of social and technical elements led to a minimum of movements and a work load producing to the highest daily output: for example, Gilbreth, an associate of Taylor, designed work procedures and scaffolds for bricklayers that led to the reduction of movements from 18 to 5 and tripling the number from bricks laid per hour from 120 to 350 (Taylor, 1911, pp. 37–40). Taylor also stressed the importance of foremen and management teaching workers if they were not productive enough, though the piece-rate system was more effective: pay increases in proportion to the quantity of output. Taylor did not use the term socio-technical, but clearly scientific management addressed how functions and responsibilities were to be distributed between technology, work procedures, and organization through empirical studies of workers through scientific management and decisions by management.

The interest into empirical studies of work, technology, and organization to achieve maximum efficiency was perpetuated in the so-called Hawthorne Studies in the USA between 1924 and 1933 where breaks, daylight, pay increases, free meals, and other elements were systematically varied to find out which combination would lead to maximum productivity (Mayo, 1933). One of the surprising outcomes, which have made these studies well-known in the history of organizational theory, was that the recognition and acknowledgment of workers as persons by the observing sociologist, and not just regarding workers as mere cogs-in-the-wheel of the production facility, which was the approach of scientific management, led to the highest productivity at all. With this, the "Hawthorn effect" became part of organizational theory as the insight that the observer influences the observed, but also more broadly that what we could call a "rationale of recognition" can positively influence employees' productivity.

However, it was the Tavistock research group in the 1950s in the United Kingdom that coined the term "socio-technical system" from their studies of coal-mining (Trist and Bamforth, 1951). Contrary to Taylor's focus on the individual worker and singular tasks, they viewed employees as complementary to rather than extensions of machines; focused on the group as the central work unit and argued for letting groups regulate work internally rather than by external managerial control. Also, whereas Taylor argued that there was one optimal, scientific way to organize work, the Tavistock group argued that there were multiple ways in which work could be organized optimally (what is called equifinality), and that the aim was to strive for an optimal match between the social—the requirement of the individual worker, the group, and the organization—and the technical. In this match, efficiency of organization as well as work quality and groups as semiautonomous learning systems was emphasized (Trist, 1981). By implication, employees might be as good at finding the optimal fit as scientists or management. For lack of a better word, we could label this the "socio-technical" design rationale.

The aim here is not to go through the entire development of organizational theory, but merely to point to the fact that the interweaving of social and technical elements have been discussed for more than 100 years, and that different design rationales stress different aspects of how work can and should be organized through and with technology. The different design rationales also imply different approaches to how design of work and technology should be conducted, while all emphasizing empirical studies of work practices: scientific management emphasizes the delegation of decision-making upon the design of work practices to scientists and management; the Hawthorne studies point to the importance of recognition and appreciation; whereas the socio-technical approach of the Tavistock group points to the importance of including employees in the design. Hence, the betwixtness of design and the need to choose between different design rationales has for over 100 years been an inherent challenge of designing work, technologies, and organizations. However, often this betwixtness is covered over by claims of "best practice", "science", or a new trendy organizational management approach. While such claims may serve to sell the message of a new design, they are rarely a good starting point of the actual design process itself, since being reflexive about how to balance different rationales and being explicit about choices works better for the long-term outcome of the socio-technical system's success.

While the above has focused on different kinds of industrial production, the same design rationales have also been applied for health care organizations and technologies. Thus Gilbreth as proponent of scientific management in 1914 argued that "… the same laws which govern efficient shop practices also govern efficient practice in the hospital, … [and] … many of the problems involved are not only similar, but identical, and the many of the solution which we have found to those problems in the shop can be carried over bodily into the field of hospital" (Gilbreth, 1914; For the transfer of scientific management and the Total Quality Movement from business to health care, see Arndt and Bigelow, 2015). Gilbreth mainly had transportation and surgery in mind, and scientific management has had limited impact on professions such as doctors and nurses, because their work relies on the adequate application of expertise and choice of action, and hence is difficult to break down in separate tasks and steps to be planned by external management (for the failure of scientific management in the case of surgery, see Whitfield, 2015.) However, the professional autonomy of doctors and nurses makes a rationale of "appreciation and recognition" more adequate. As for the socio-technical rationale, it has been argued that hospitals are complex organization where exception to routine and procedure is so prevalent that they require nuanced recognition of the interrelations of people, work, and technology (Lorenzi et al., 1997). An argument that has been made in connection with efforts to develop information systems in general and in relation to health care IT in particular (e.g., Baxter and Sommerville, 2011; Berg, 1999; Whetton, 2005).

The existence of different rationales and approaches to the design of the interrelations of people, organization, and technology positions developers of health care IT in a betwixt position. This has not only to do with the existence of different value systems within socio-technical approaches where a humanistic approach emphasizes work life quality and employee satisfaction may be opposed to a managerial approach emphasizing the company objectives which most often pivot around economic issues. In the best scenarios, the first leads to improved productivity and all is good, whereas in other cases conflicts arise (Baxter and Sommerville, 2011, p. 8). But the issue is complicated by the different rationales outlined above cutting across value systems and concerns also how work practices are conceived and shaped at a more fundamental level as mentioned in connection with the second betwixtness: the existence of multiple perspectives upon how actors, work, and organizations do or should function, does not only posit designers or system developers in position where they "just" have to choose between different value systems. They also have to investigate how work should be conceptualized.

In the following, I will present and analyze two cases of design and use of health care IT to provide concrete illustrations of how the kinds of complications and betwixtness may unfold. This will provide the material for the discussion in the last section.

3. A FOUNDATIONAL MODEL FOR ELECTRONIC HEALTH RECORDS

Visions of computerized patient records have been launched since the 1960s, but, as stated in a review in 2005 in the Journal of American Medical Informatics Association: despite 30 years of predictions that the use of EHRs would soon become widespread, "the wave has never broken": Technology immaturity, health administrator focus on financial systems, lack of application friendliness, and physician resistance had been barriers to such a development (Berner et al., 2005). However, the general spread of personal computers and computer skills, pressures for higher efficiency in health care through IT, and the spread of scientific thinking, including standardization, within the medical professions converged to make EHRs more likely to become realized. Indeed, already in 1968 the physician Weed had proposed a problem-oriented medical record (POMR) as a structured, more scientific alternative to the narrative-based documentation and presentation of patient cases. This was possible with paper-based records, but the full potential POMRs would only be realized with computers, Weed argued (Weed, 1969). Overall, the ground for making the wave break was laid.

Thus, the first decade in the 2000s did see several major efforts to develop EHRs in Western countries, including the USA, Canada, the UK, France, and Norway. This was also the case in Denmark in which the following case unfolded, and where the development of EHRs was a major issue for shifting governments, who urged the regions, which are in charge of hospitals, to develop or procure EHRs. However, there was concern that such EHRs might not be interoperable between regions or able to exchange data upon patients, which might pose a problem for continuity of treatment and care, if citizens moved from one region to another. To avoid this, the National Board of Health in Denmark engaged in a project to develop a "basic model for EHRs" (BEHR) to which all EHRs developed or procured by the regions had to comply. In addition to information exchange, there were two further purposes: To ensure that information on diseases and treatments could be reported to a national database, and to create cross-professional EHRs. The former was necessary for governing the welfare state–based and financed health care sector, while the latter was desired to make, for example, physicians and nurses coordinate and work closer together when treating and providing care for patients.

BEHR was based on a model of how humans in general were perceived to approach problems: first, a problem is considered; then a plan for action is made; subsequently the plan is executed; and finally the outcome is evaluated. This putative, general problem-solving model was applied to health care as "Clinical Process" and depicted in the following way (Fig. 5.2):

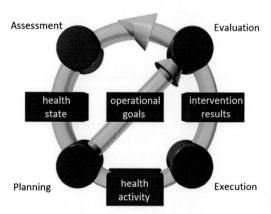

FIGURE 5.2 Clinical process, the central conceptual core of BEHR.[1] *BEHR*, basic model for electronic health records. *Available from https://www.sst.dk/applikationer/epj/gepj/022_20050520/index.html.*

[1] Translated into English from: "Beskrivelse af GEPJ – på begrebsniveau" ("Description of BEHR – at conceptual level").

Clinical problem solving by doctors, nurses, and other health care professionals, was hence seen as involving four steps: The health care professional obtains information upon a patient's health state; plans for a set of health activities; these are then executed producing a number of intervention results, which, finally, are evaluated, and a new cycle of problem solving can be initiated. Clinical process was argued by the National Board of Health to depict the work of health care professionals on a mental, conceptual, and computer science level. Mentally, it allegedly depicted how clinicians thought, conceptually it provided a model for how clinical information should be structured, and computationally it provided framework for how information should be modeled (details of this case can be found in Bossen, 2006a,b).

Since, as mentioned previously, representations are always made from a specific position and context, a first objection to this linear representation might by that it reflects an administrative, desk-top made view of clinical practice removed from actual practice, rather than from empirical insights and practical experience. If this was the case, we might most likely expect BEHR to fail. However, the model was actually based on the physician Weed's suggestion for a problem-oriented medical record mentioned previously (Weed, 1968, 1969). Also, BEHR was presented, developed, and refined by the National Health Board in cooperation with doctors and nurses, who at meetings approved BEHR as depicting how they worked: the description and representation of the work of physicians and nurses was thus made from a position within these professions and within a health care context. Finally, the National Board of Health wanted to test the model in use, and a number of working prototypes were developed and pilot tested at hospital departments, which, despite expected problems such as software bugs, implementation issues, and necessary changes to work practices, did conclude that BEHR was a viable model.

However, an instructive controversy arose in connection with the last and largest of these tests that lasted 4 months during with 66 patients had their treatment and care documented through a newly developed, BEHR-based prototype: the subsequent evaluation argued that BEHR was *not* adequate for clinical work. If true, this was a major blow to the effort of the National Board of Health, but of particular interest here, is the issue of whether BEHR indeed did describe and represent clinical work adequately, at heart of the second kind of betwixtness outlined above.

The evaluation report was based on ethnographic observations and interviews and pointed to three problems: First, the report argued that the BEHR model entailed a fragmentation of patient cases, since it demanded that each health care problem to be entered separately with each its own cycle of planning, intervention, results, and evaluation. Whereas this worked for patients with only one problem, patients with multiple health problems would have documented these separately, which meant that the interactions and interrelations between these were difficult to get hold of. Second, clinicians argued that they lost overview of patient information due to the previously mentioned fragmentation of patients into separate health problems, and because information could be put at several places. For example, an X-ray of the thorax is relevant for pneumonia as well as for cardiac problems, and the doctors were uncertain as to whether they should put the information in one or all relevant health problems: the first solution meant that such information might be overseen even though relevant for more health problems, while the second solution meant repetitive and an overflow of information.

The more health problems, the more the challenges of documentation and finding information were exacerbated. Third, doctors and nurses argued that the BEHR prototype entailed more work, since it required them to document more information and broke documentation down to separate pieces.

In the light of this criticism, an argument arose as to whether the problems were caused by the specific design of the prototype or by the BEHR model itself. The National Board of Health argued that the prototype had implemented the model too rigidly in the user interface and that the model was meant to work at the back (See Fig. 5.3, below). The implication of this argument seemed to be that BEHR adequately depicted clinical decision-making at the mental, computational, and informational levels, but not at the level of practical use.

Others, including this author, argued that the problems were to be attributed to the model itself: BEHR depicts clinicians viewed outside of situated practice as problem-solving individuals whose work is that of producing rational accounts of patient cases through sequenced steps. Essentially, the general problem-solving model and BEHR align with the linear, rationalistic depiction of work outlined above (Fig. 5.1, Top). In this sense, the BEHR model is cognitivist. What BEHR misrepresents is the iterative, cooperative process that finding, filtering, valuing, and connecting information which is involved when

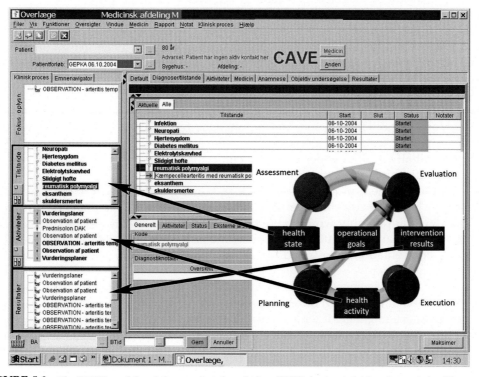

FIGURE 5.3   The BEHR model and the user interface design. *BEHR*, basic model for electronic health records.

clinicians reason about patient cases within practice. When looking closely, clinical work concurs much closer with the interactional depiction of work (Fig. 5.1, Bottom).

An alternative approach to clinical reasoning closer to the interactional depiction can be found in, for example, the work of Feinstein (Feinstein, 1973a,b,c). Here, clinical reasoning is seen as a matter of finding, assessing, and filtering information with the goal of constructing a coherent account of the patient's situation. This involves two main phases: in the first, the clinician gathers, assesses, and filters various pieces of information with the aim of constructing a coherent account; in the second, a rational account based on the previous process is given and where previously fragmented information is made into a coherent account. While BEHR fits to the latter phase of documenting rational accounts, it does not fit with the iterative, experimental process of the former. Like the computerized patient record of general practitioners in the UK, investigated by Heath and Luff (1996), BEHR can be said to be "a disembodied, retrospective account of the consultation, rather than an integral feature of the accomplishment of diagnostic and prognostic activities" (Heath and Luff, 1996, p. 363). Obviously, the argument here is not that clinical reasoning should not be rational, but rather that BEHR only partially describes and represents the unfolding of such reasoning. It may represent how clinicians think—who knows?—and the genre of making rational accounts, but not the actual practice of finding, assessing, and filtering the information. It also focuses on the individual clinician rather than on the network of people and artifacts through which clinical work, including diagnosing, is actually achieved (Atkinson, 1995).

One of the challenges of BEHR then is the linear, rationalistic, and individualist design rationale behind it, and the way it depicts practice, which is contrary to more ecological and practice-oriented accounts of clinical work (Atkinson, 1995; Jensen and Bossen, 2016). The challenge for designers and developers of IT in this case is not that of choosing between different value systems, nor that of whether to involve workers or users (i.e., clinicians) in the development of IT or not. BEHR was intended to support the health care management and clinical work at the same time, and clinicians were involved in its development. However, BEHR adhered to a rationalistic representation of clinical work that was only adequate for part of actual practice. Further, it is very possible that the BEHR model was initially positively perceived by clinicians, because it resonated well with the kind of scientific thinking in which doctors are trained, and which also has strong proponents within other medical professions such as nursing and midwifery (Clausen, 2010). However, the latter are also characterized by hermeneutical and phenomenological forms of knowledge that probably would have been difficult to adjust to BEHR's demands for documentation (Clausen, 2010). In that sense, the issue around BEHR is both about how practice is represented and designed for, as well as about which kinds of knowledge it supports.

The case of BEHR demonstrates the intrinsic betwixtness of doing socio-technical design by pointing out that even when different perspectives and stakeholders are involved in the design process, one might get it wrong. Here, the misrepresentation did not show up before the prototype was used in actual practice. In the end, BEHR was abandoned in the aftermath of the controversies following the last prototype test described above and disappeared from the National Board of Health's strategic plans. Instead of a national basic model for EHRs, each region was free to develop or procure the EHRs it found fit.

4. EMPOWERING AND MANAGING DISTRIBUTED WORK: IT FOR HOSPITAL PORTERS

In the first case, the focus was upon the complexities involved in representing the work that the IT system was designed to support or change. In the second case below focusing on hospital porters' work, the issue will be more upon the choice between different design rationales.

It involves an IT system developed to support hospital logistics by tracking equipment, beds, and vehicles transporting linen, medicine, and food via RFID and thus enabling the location of these and coordinating requests for them. This includes hospital porter services of moving patients in and out of beds, transport of patients between departments and to and from examinations in wheelchairs and beds—most often X-ray and MR-scanning—and moving equipment around. Despite being central to the functioning of hospitals, porters are relatively invisible within research, and often their work is described from an operations research perspective focusing on how to balance demand for services with the number of porters: there should be enough porters available at peak times, but not too many idle at times with less load (i.e., the linear, rationalistic depiction of work, Fig. 5.1, Top). The problem to be solved is most often conceived of the challenges of "work load" and porters' work is seen more or less from a rationale of scientific management (notable exceptions are Odegaard et al., 2007; Rapport, 2009).

The IT system in question was developed in the early 2010s in a process involving hospital porters and local management, and implemented at a university hospital, which has approximately 6500 employees, of which 114 are porters, and which has 960 beds and treats more than 550,000 outpatients every year (For details, see Bossen and Foss, 2016; Stisen et al., 2016). Before the implementation of this Task Management System (TMS), clinical staff would call a central dispatcher on phone whenever they needed a porter. Receiving the call, the dispatcher would note down necessary information such as time, place, level of urgency, necessary equipment or means of transport, and subsequently either delegate the task to one of the porters present in the Porter Central, or make a phone call to a porter whom the dispatcher based on his year-long experience assessed to be available or close to where the task was to begin. To get an overview, the dispatcher had the names of available porters on small wooden blocks, which he moved between two zones indicating "available" or "on assignment", as well as a list of incoming on paper request that was continuously updated. While this system worked, it had some disadvantages: misunderstandings were frequent, because information was passed on verbally; tasks were not always given to the porter closest to a task, but to the first one available though far away and thus ultimately late at arriving; and it was difficult to handle the incoming stream of ad-hoc tasks centrally with the existing technology (pen, paper, wooden blocks with porter names, and a line dividing a table surface into two zones of "available" and "on assignment"). Hence, the developed TMS was to create a more efficient workflow and reduce waiting times, which sometimes resulted in canceled treatments and surgeries.

The TMS allows clinicians to enter requests for porter services digitally, and these requests are then visible to porters on smartphones, including essential information on where, when, level of urgency, necessary equipment, and so forth. The smartphones display a list of all requests not yet assigned as well as the location of porters, and this information is also available centrally to the dispatcher and porters at the Porter Central. Porters themselves pick their next task, and because their locations are tracked via the smartphones and made visible, they can see who is available and closest to the tasks requested. While the dispatcher is not necessary with the new system, the hospital decided to keep this function, and rather than coordinate requests his job has changed into communicating with staff at the departments solving technical issues with the new IT system, dealing with inquiries upon long waiting times, as well as overseeing that unpopular requests are not ignored by porters.

The TMS entailed a number of changes to the work of porters and departments too. The departments had screens showing all tasks they had requested and whether they had been taken by a porter or not, and could ensure that information was not lost. Porters could see all requests by all departments, their urgency, and the location and availability of their colleagues, and most importantly could choose tasks themselves. They could use their own experience about how to do tasks most efficiently depending on the nature of the task and their current location, and hence contribute to the overall coordination of porter services. They had more influence upon and took more responsibility for their work. The system increased the porters' overall awareness of the overall work load upon porter services as well as upon the work of their colleagues. Thus, as one porter said,

> Previously, I had a bit of this feeling of "am I the only one making an effort?" At that time, the phone called all the time, and when you had just ended a call, the next would come in. I was stressed. And then you start wondering whether the others just sit and drink coffee. But now I can see that they are working as much as I am. And I just take one task at a time, and am aware of what is most pressing. *Male Porter, 34 years.*

At the same time, the list of requests seemed at times never ending, and some porters felt more pressured, because whenever they had completed one task, there would be many others waiting on the list. For the same reason, they decided not to have the list displayed at the large screen in the room where they had their breaks, as they could not relax with new requests continuously coming in. Notably, the 35 porters on average on a normal day handle around 700 requests with peak hours around noon, when ward rounds are completed and new examinations and treatments have been ordered. Overall though, porters appreciated the increase of influence upon and responsibility for their job. The system and the way it was implemented supported an interactional model of work (Fig. 5.1, Bottom), and coordinating work from "within" the complexities of situated action (Bowers et al., 1995). At this stage then, the design rationale behind the new system seems close to that of "socio-technical design" where a match between organizational, technological, and work group requirements has to be found.

In addition to the TMS, a Porter Management Information system was developed which built upon the data from the former and enabled accumulating and calculating data such as the number of tasks per department, the kinds of tasks completed per hour, day and week, as well as average waiting and response times. The Porter Management Information system was intended to provide hospital management information upon the porters' performances and to pinpoint inefficiencies in the organization of hospital logistics, and to provide the dispatcher information so as to, for example, compare the request rhythms of different departments to the rhythm of the overall work load. With such information, he could explain why, for example, there were waiting times at certain departments or certain hours, and he gained insights into porters' work and could optimize it. During our study, this worked well: the dispatcher used the information to argue for hiring more porters, since he could show that there was not sufficient staff to cover peak hours. Also, he could argue and show one department that long waiting times were due to repairs to elevators which forced bed transports to wait for an available elevator. To another department, he could suggest to change when to order requests, because these peaked at the same time as the overall load peaked. Further, he could counter porters' tendency to ignore requests from a far-away department. As the latter examples suggests, the Porter Management Information system could be used to monitor the work of porters as a group and individually. This made the porters somewhat uncomfortable, but at this point management did not want to pursue this course, because they thought that more was gained by trusting porters to take responsibility for their work in the spirit of a rationale of recognition. Rather, management was concerned about internal competition and bullying, since the porters could monitor their colleagues more closely now.

As this second case about designing for hospital porters work shows, IT systems often find themselves at the crossroad of several intersecting interests and uses and can be used for control and accountability as well as for enhancing work life quality. This in turn is also related to how porters' work is perceived and represented. Regarding the issue of different usages, it can be noted that in addition to improved communication between departments and porters and reduction of misunderstandings, the TMS also enables the coordination of work by porters themselves and allows them to gain more influence upon and responsibility for their work. This is much in line with the rationale of socio-technical design. Indeed, the system also allows for the function of the dispatcher to go away and leave the coordination of porter services to the porters themselves as a group going even further in the direction argued for by the Tavistock group. At the same time, the associated Porter Information System enables close monitoring of porters as a group and as individuals, which can be used to adjust the number of staff to the rhythm of work load during the day, argue for more staff, and point at inefficiencies (e.g., lack of elevators) in hospital logistics. However, it can also be used to monitor the efficiency of individual porters and linked to pay incentives. Such a move toward a rationale of scientific management would be facilitated by hospital porters' position at the bottom of the hospital hierarchy, and the prevalent perception that their work is not "skilled" or "knowledge work" as, for example physicians' and nurses' work. As Rapport's ethnographic study shows, this is partially a misrepresentation, but nonetheless the approach of much literature upon porters' work (Rapport, 2009; Bossen and Foss, 2016). Usage of an IT system is thus related to the issue of perceptions of work. Already, part of porters' micro-coordination is made invisible by the system, since it does not allow for documentation of handling tasks in parallel. For example, a porter might be on the way to get a patient out of bed, but picks up a wheelchair in the way, because the next task is a transport where this is needed. Overall time is saved, but the coordinative effort not visible. This is the kind of invisibility produced when work tasks are depicted in linear, rationalistic ways, since the interactional efforts involved are glossed over (as visualized by the difference between Fig. 5. 1 Top and Bottom). Also, the beginning and completion of tasks in the system were defined by when porters accepted requests and by when they arrived at departments with equipment or patients and recorded this. However, before and after a patient transport, a porter might have to get a wheel chair or other equipment from a depot, and bring it back again. With the present practice of accepting and completing tasks this time would show up the system as idle, which was of concern to porters when reflecting on how management might use the information from the systems in the future. Efficiency efforts might then be ill-directed if based on the numbers and assessment of work load and porter efficiency produced through data in the two systems only.

5. DISCUSSION: ON THE BETWIXNESS OF DESIGN IN HEALTH CARE

A socio-technical approach to the design of new health care IT has the advantage that it foreground two issues from the start: First, the task of distributing competences, responsibilities, and tasks between the social and the technical, and second the importance of paying attention to the people who will take the system into use. The design challenge is intrinsically social and technical at the same time. These issues are especially pertinent in health care for a number of reasons that make the socio-technical intricacies especially pronounced: knowledge work is central and makes automation difficult; the role of professions such as doctors and nurses is strong making any change without their involvement tough; the mutual and interrelated development of technologies and expertise is already high; work is contingent and dynamic demanding flexible technologies and organizational set-ups; it is a high-risk, high-safety domain; and patients, staff, management, politicians, and the public form strong interest groups that have to be accommodated (Glouberman and Mintzberg, 2001b, 2001a).

Central principles in socio-technical design are "to put owners of problems in charge of design and use" (Fischer and Giaccardi, 2006), and that "design should reflect the needs of the business, its users, and their managers" (Clegg, 2000). The strength of these principles is that it encourages the inclusion, for example, of patients, physicians, nurses and other health care staff, as well as health care authorities and managers when engaging the development of new technology. However, one challenge is to decide who are the "owners of the problem" and relevant stakeholders, and subsequently establish processes and fora in which different interpretations and solutions to the problem can be presented and discussed. EHRs for example are of central interest to multiple stakeholders: to physicians and nurses, since it is through EHRs that they coordinate and document their work; to patients, since this is where their diagnoses, treatment, and care is documented, and to health care authorities as legal documents in cases of malpractice, and as the basis of accumulating information upon the overall use of medicine and delivered health care services. In the case of BEHR, health care professions and authorities were involved, but no patients, since these were not, at this early stage, considered being "problem owners". On the other hand, while hospital porters may analytically clearly seem to be problem owners in the case of the TMS, it will be up to negotiations between them and management whether they will be included in developing the socio-technical systems around them. Here the choice will be between different rationales of design as outlined above. While proponents of socio-technical design in the Tavistock tradition argue for setting up processes that facilitate the inclusion of various stakeholders (Clegg, 2000; Fischer and Herrmann, 2011; Trist, 1981), actually doing so is much harder. However, a strong argument for a socio-technical approach in development phase is that technologies that fit work practices and are acceptable or even promoted by users will make the organization more efficient. Similarly, a strong argument for a socio-technical approach in the phase after implementation is that change management, adjustment of the organization to the new technology, and benefits realization is most effective when including affected stakeholders.

The issues of socio-technical distribution of competences and functions and user involvement are by themselves by no means trivial, and this chapter has focused on two challenges betwixing within socio-technical design: the challenges of generating an adequate description of the existing socio-technical work practices which the new system is going to change, and the challenge of aligning or choosing between different rationales when designing the new socio-technical systems. The chapter analyzed this betwixtness of socio-technical design through the case of a foundational model for EHRs and the case of an IT system for hospital porters' work. The first case showed that even when the future users of a system become part of a socio-technical design process, the challenge of generating an adequate description of the work practices can be difficult. Thus in the case of BEHR, the inadequacy of a linear, rationalistic depiction of clinicians work did not become apparent before it was used in ways close to everyday practice. The depiction of work close to the rational of scientific management represented work from without and did not capture the intricacies of interactional practices of clinical work. In the case of the IT system for hospital porter on the other hand, the strength of allowing porters to unfold their interactional practices and manage these on their own led to a better quality of work life, though with potentially more stress and the risk of more managerial control and surveillance mixed in. However, at the early stage of implementation at least this was a case where a socio-technical and an appreciative rationale were supported by the socio-technical design. While new technologies offer various opportunities for more control by management, they also offer opportunities for letting employees themselves coordinate and ensure quality and efficiency. In contrast to BEHR into which a depiction of work practices was strongly embedded in the design and codes itself, the hospital porter system's design was sufficiently open, or "underdesigned" (Fischer and Herrmann, 2011), to allow for a variety of different usages and redesign after implementation.

In general, such under or flexible design is important in order for employees and management to work out the best fit between work, IT system, and organization and thus work toward an optimal fit. Experience shows that such fit can be achieved in different ways (the principle of equifinality), and therefore it is crucial that a process of learning and negotiation is incorporated also after implementation. The process of "design" should thus not only be conceptualized as a phase leading toward the development and deployment of technology only, but also as a process that continues after the technology has been taken into use (Clegg, 2000; Fischer and Herrmann, 2011). Such an approach does opens up for a greater degree of complexity in the design process from the beginning to the end. It seems simpler and more efficient just to go straight for the most efficient technological solution first, and then design work practices around it, or vice versa, but numerous example show that it is not (Bowers et al., 1995; Trist and Bamforth, 1951): Because the question of how to distribute functions and roles between humans and technologies is at the heart of the complexity, and because answering that question necessitates the inclusion of various stakeholders, the design process must be a prolonged process of learning and redesign.

Choices in how to approach design might make certain group silent in the process (e.g., patients, medical secretaries, or nurses), but a profoundly analytical or perceptual challenge is when certain work process are made silent as the case of BEHR also shows. The inclusion of various stakeholders and perspectives from various problem owners is crucial for the design of workable systems, but does not ensure that the work practice itself is conceived adequately. Looking at work from outside of practice may lead to a rationalistic view of that work practice, thus missing core coordinative and cooperative elements. The answers to this challenge are detailed empirical studies, prototyping, and pilot tests of technology so that it may unfold whether and how the technical fits the social. Socio-technical design is in the broadest sense as a distribution of function roles between humans and machines unavoidable, and in the Tavistock tradition as a drive for an optimal fit between optimal social and technical systems a great vision to strive for. At the same time, socio-technical design by necessity and unavoidably involves a complexity and betwixtness of design that needs attention.

6. CENTRAL POINTS

- Socio-technical design is about distributing competences, functions, and tasks between people, technology, and organization
- Socio-technical design provides a strong argument for including affected people and stakeholder in the design process
- Socio-technical design involves choosing between different rationales for design reflecting different stances toward and interests in work and organizing (e.g., scientific management, appreciation, socio-technical rationales)
- Socio-technical design involves the nontrivial task of generating an adequate depiction of work practices (e.g., a linear, rationalistic, or an interactional model)
- Socio-technical design takes places before and after the implementation of an IT system
- Socio-technical design in health care is especially relevant and difficult because of the domain is intrinsically shaped by an interwovenness of technology and expertise, because of the importance of knowledge work, and because of the number of strong stakeholders.

Acknowledgments

The author thanks Martin Foss for the cooperation upon hospital porters, and permission to use that case in this chapter.

References

Arndt, M., Bigelow, B., 2015. The transfer of business practices into hospitals: history and implications. Advances in Health Care Management 1, 339–368.

Atkinson, P., 1995. Medical Talk and Medical Work. Sage, London.

Baxter, G., Sommerville, I., 2011. Socio-technical systems: from design methods to systems engineering. Interacting With Computers 23 (1), 4–17.

Berg, M., 1999. Patient care information systems and health care work: a sociotechnical approach. International Journal of Medical Informatics 55 (2), 87–101.

Berner, E.S., Detmer, D.E., Simborg, D., 2005. Will the wave finally break? A brief view of the adoption of electronic medical records in the United States. Journal of American Medical Informatics Association 12 (1), 3–7.

Bossen, C., 2006a. Participation, power, critique: constructing a standard for electronic patient records. In: Proceedings of the 9th Conference on Participatory Design, pp. 95–104.

Bossen, C., 2006b. Representations at work: a national standard for electronic health records. In: Proceedings of the 2006 Conference on Computer Supported Cooperative Work, pp. 69–78.

Bossen, C., Foss, M., 2016. The collaborative work of hospital porters: accountability, visibility and configurations of work. In: Proceedings of the 19th ACM Conference on Computer-Supported Cooperative Work & Social Computing, pp. 965–979.

Bowers, J., Button, G., Sharrock, W., 1995. Workflow from within and without: technology and cooperative work on the print industry shopfloor. In: Proceedings of the Fourth European Conference on Computer-Supported Cooperative Work, pp. 51–66.

Clausen, J.A., 2010. How Does Materiality Shape Childbirth? An Explorative Journey into Evidence, Childbirth Practices and Science and Technology Studies (Unpublished Ph.D. thesis). Aarhus University.

Clegg, C.W., 2000. Sociotechnical principles for system design. Applied Ergonomics 31 (5), 463–477.

Eskerod, P., Huemann, M., Savage, G., 2015. Project stakeholder management—past and present. Project Management Journal 46 (6), 6–14.

Feinstein, A.R., 1973a. An analysis of diagnostic reasoning. I. The domains and disorders of clinical macrobiology. Yale Journal of Biology and Medicine 46, 212–232.

Feinstein, A.R., 1973b. An analysis of diagnostic reasoning. II. The strategy of intermediate decisions. Yale Journal of Biology and Medicine 46, 264–283.

Feinstein, A.R., 1973c. The problems of the "Problem-oriented medical record". Annals of Internal Medicine 78 (5), 751–762.

Fischer, G., Giaccardi, E., 2006. Meta-design: a framework for the future of end user development. In: Lieberman, H., Paternò, F., Wulf, V. (Eds.), End User Development – Empowering People to Flexibly Employ Advanced Information and Communication Technology. Klüwer, Dordrecht, 427–457.

Fischer, G., Herrmann, T., 2011. Socio-technical systems: a meta-design perspective. International Journal of Sociotechnology and Knowledge Development 3 (1), 1–33.

Gilbreth, F.B., 1914. Scientific management in the hospital. Modern Hospital 3, 321–324.

Glouberman, S., Mintzberg, H., 2001a. Managing the care of health and the cure of disease – Part II: integration. Health Care Management Review 70–84.

Glouberman, S., Mintzberg, H., 2001b. Managing the care of health and the cure of disease – Part I: differentiation. Health Care Management Review 56–69.

Greenhalgh, T., Potts, H.W., Wong, G., Bark, P., Swinglehurst, D., 2009. Tensions and paradoxes in electronic patient record research: a systematic literature review using the meta-narrative method. Milbank Quarterly 87 (4), 729–788.

Heath, C., Luff, P., 1996. Documents and professional practice: 'bad' organisational reasons for 'good' clinical records. In: Ackerman, M.S. (Ed.), Proceedings of the 1996 ACM Conference on Computer Supported Cooperative Work. ACM Press, New York, pp. 354–363.

Jensen, L.G., Bossen, C., 2016. Distributed plot-making – creating an overview via paper-based and electronic patient records. Scandinavian Journal of Information Systems 28 (2), 3–27.

Latour, B., 1992. Where are the missing masses? The sociology of a few mundane artifacts. In: Bijker, W.E., Law, J. (Eds.), Shaping Technology/Building Society. MIT Press, Cambridge (Mass), London, pp. 225–258.

Law, J., 1992. Notes on the theory of the actor-network: ordering, strategy, and heterogeneity. Systems Practice 5 (4), 379–393.

Lorenzi, N.M., Riley, R.T., Blyth, A.J., Southon, G., Dixon, B.J., 1997. Antecedents of the people and organizational aspects of medical informatics: review of the literature. Journal of the American Medical Informatics Association 4 (2), 79–93.

Markus, L., 1983. Power, politics, and MIS implementation. Communications of the ACM 26 (6), 430–444.

Mayo, E., 1933. The Human Problems of an Industrial Civilization. MacMillan, New York.

Odegaard, F., Chen, L., Quee, R., Puterman, M.L., 2007. Improving the efficiency of hospital porter services, Part 1: Study objectives and results. Journal for Healthcare Quality 29 (1), 4–11.

Rapport, N., 2009. Of Orderlies and Men: Hospital Porters Achieving Wellness at Work. Carolina Academic Press, Durham.

Simonsen, J., Robertson, T. (Eds.), 2012. Routledge International Handbook of Participatory Design. Routledge, Oxford.

Stisen, A., Verdezoto, N., Blunck, H., Baun Kjærgaard, M., Grønbæk, K., 2016. Accounting for the invisible work of hospital orderlies: designing for local and global coordination. In: Proceedings of the 19th ACM Conference on Computer-Supported Cooperative Work & Social Computing, pp. 980–992.

Strauss, A., 1993. The Continual Permutations of Action. Aldine de Gruyter, New York.

Suchman, L., 1995. Making work visible. Communications of the ACM 38 (9), 56–63.

Taylor, F.W., 1911. The Principles of Scientific Management. Harper & Brothers, New York, London.

Trist, E.L., 1981. The Evolution of Socio-Technical Systems: a Conceptual Framework and an Action Research Program Occasional Paper. Ontario Quality of Working Life Centre. 2.

Trist, E.L., Bamforth, K.W., 1951. Some social and psychological consequences of the longwall method of coal-getting. Human Relations 41 (1), 3–38.

Weed, L.L., 1968. Medical records that guide and teach. New England Journal of Medicine 278 (11 + 12) 593–600 + 652–597.

Weed, L.L., 1969. Medical Records, Medical Education, and Patient Care. The Problem-Oriented Record as a Basic Tool. Year Book Medical Publishers, Chicago.

Whetton, S., 2005. Health Informatics: a Socio-Technical Perspective. Oxford University Press.

Whitfield, N., 2015. Surgical skills beyond scientific management. Medical History 59 (3), 421–442.

Stakeholders as Mindful Designers: Adjusting Capabilities Rather Than Needs in Computer-Supported Daily Workforce Planning

Martina Augl[1], Christian Stary[2]
[1]Tirol Kliniken, Innsbruck, Austria; [2]University of Linz, Linz, Austria

1. INTRODUCTION

Running a hospital and its organizational units, in particular those related to patient care, requires coordination and cooperation among stakeholders and across functional hierarchies. Stakeholders are persons, communities, or organizational units involved in work procedures or business process. In the course of accomplishing (work) tasks, in most cases, they interact with technology thus becoming users and adopters of information and communications technologies. In this way, stakeholders affect others or are affected by their activities, either directly or indirectly.

Providing facilities for stationary and ambulatory patient care require the adjustment of various resources for daily work. The capacities and needs of doctors, nurses, technicians, administrators, and patients must be recognized and adjusted for patient-oriented operation based on various IT support systems. The alignment of available competences to accomplish tasks through the help of IT support systems is part of the process of designing socio-technical systems. It affects various stakeholders, such as operation planners and patients, and also affects technologies supporting health care procedures, such as planning the availability of expert teams for case-sensitive, nonstationary treatment of patients. Stakeholders and technology interrelate and affect each other. The design of socio-technical systems needs to reflect these effects in a balanced and structured way in the course of developing organizations.

Work improvement based on business process models has been recognized as an enabler of technical and organizational development and thus as a critical success factor for health care organizations worldwide (Rebuge and Ferreira, 2012). Thus, process and technical system (re)design is considered essential, aiming to uncover the fundamental needs underlying the structuring of health care organizations. The needs of stakeholders and their articulated potential for change should trigger further development steps. These steps are typically addressed primarily from a cost or effort perspective (Gershengorn et al., 2014) and justified by the increasing financial pressure on the health care sector and its institutions (Anyanwu et al., 2003).

While implementing measurement systems to meet economic objectives, concerned stakeholders are increasingly participating in the improvement of core processes, such as workforce planning, particularly when patient orientation, and continuous dynamic adaptation of technology have become essential parts of health care system development (Lenz and Kuhn, 2004). Involving stakeholders is expected to lead to user-centered information systems and effectively supported workflows (Ghazali et al., 2014; Zhang, 2005).

Planning an expert workforce is critical, both for stationary and walk-in care, since it affects the quality of patient care and efficiency of resource sharing (Augl and Stary, 2015). Still, planning support requires conceptual and technical rethinking since existing approaches lack methodological maturity, such as transparent mapping of needs to features, hindering stakeholder acceptance (Lopes et al., 2015). Although technological support helps in identifying process knowledge, for example, during process mining (Rebuge and Ferreira, 2012), stakeholders must be part of socio-technical design in order to provide organizations with the knowledge how their processes are currently being performed. This knowledge can then be used to develop alternative work behaviors, for example, due to conflicting goals, and in a way organizations can improve their processes and systems in alignment with their actual capabilities (Prilla et al., 2012). Such a procedure helps to avoid follow-up efforts and costs, once certain support systems have been introduced to the work place.

Like previous approaches, we followed the principles of participatory action research (Robert, 2013). We also drew on experience-based codesign (Donetto et al., 2015) requiring stakeholder engagement on a social level (Trebble et al., 2014). The latter helps to overcome organizational barriers (Wong et al., 2011) while opening up the possibility for change. It aims to go beyond an approach to user-involved care planning that "is typically operationalized as a series of practice-based activities compliant with auditor standards. Meaningful involvement demands new patient-centered definitions of care planning quality. New organizational initiatives should validate time spent with service users and display more tangible and flexible commitments to meeting their needs" (Bee et al., 2015, p. 1).

The presented case concerns the continuous improvement of the daily workforce planning of clinic doctors for timely treatment of admitted as well as ambulatory patients. Doctors, nurses, and administrators at a highly specialized hospital and university clinic felt unhappy about the scheduling process for the daily operation of the clinic. They expressed a need to organize their work in a more dynamic way, in particular by capturing short-term changes in outpatient service provision, while maintaining if not increasing the service quality for patients. In this regard, transparent treatment scheduling was considered to be a key enabler. The hospital management, concerned about the reputation of the clinic, supported this project, even when it went beyond the organizational boundaries of the clinic.

By contrast to other (mostly outside driven) interventions, the project aimed to identify and implement internal opportunities and commitments offered by concerned stakeholders, who started to revisit their capabilities in terms of recognizing preceding and follow-up work tasks, recognizing the quality of exchanged data, and the like, and make organizational offers to other stakeholders. These were notably in terms of providing information relevant for accurate planning and communicating more accurately about planning issues. In addition the processes were modeled in such a way that communication and coordination played a key role, rather than individual task accomplishment. The resulting models could be executed directly after their specification. In this way, each stakeholder could experience how the anticipated behavior might become part of the entire planning process.

We first detail the study as it was designed and explain how major tasks could be accomplished before discussing the learnings for socio-technical system design.

2. DESCRIPTION AND COURSE OF STUDY

In the university clinic for visceral, transplant, and thoracic surgery at an Austrian health care institution, the quality of planning embedded in the daily scheduling of physicians had been recognized as insufficient to ensure satisfactory patient care and resource management. Besides nonstationary patient treatment in an outpatient department, stationary treatment, operations, and academic education needed to be coordinated and scheduled daily. Not only the effect of planning but also the current planning procedure had become a central bottleneck within the daily routine of the clinic.

2.1 Project Setup

As stakeholders had perceived a lack of transparency in communication, overhead resulted from iterative and redundant steps during the scheduling processes. A clinic working group and, later, a project team under the lead of the Organizational Development Department of the hospital's provider organization had been established. Besides the urgent need to revisit planning procedures, there was the will to tackle in this context the nationally promoted goals of the federal health care reform, namely to increase the efficiency and customer orientation of operational health care procedures.

The organizational development project was set up to improve daily workforce planning for the clinic's stationary and outpatient operation. The results should allow all involved operational stakeholders to accomplish their tasks as planned for the benefit of patients based on a commonly agreed and transparent schedule for all professional groups. In cases of change, the project should consider notification mechanisms to avoid any bottlenecks, including around the academic education provided by the physicians.

The project was established by a steering committee and involved three professional stakeholder groups from the clinic: doctors (6), caretaking staff (9), and administrators (7). It was structured in three phases:

- Analysis and piloting (year 1)
- Operational testing and development of a production concept (year 2)
- Final implementation and roll-out (starting in year 3)

The results were reported regularly to the steering committee according to the achievement of milestones. Three organizational development experts facilitated and monitored the project, one of them acting as a project leader and sitting on the steering committee.

In the following, we will detail analysis and piloting, as it represents a significant change with respect to the scope of planning and IT support. Hence, this phase is the most relevant for stakeholders who seek to become mindful designers of socio-technical systems. Analysis and piloting took the following course:

1. *Capturing and consolidating state-of-operational-affairs in daily workforce planning.* The following items were addressed:
   a. Which planning instruments, including technical support systems, exist for special treatments, walk-in patient care, mentoring, or teaching?
   b. Which data are in use for daily workforce planning?
   c. Which additional data need to be considered for daily workforce scheduling, for example, as already demanded by various stakeholders or as evident from experience with existing planning instruments?
2. *Communication analysis of the actual planning procedure* by exploring interaction patterns among stakeholders, including the information they exchange.
3. *Participatory design and prototyping of a technical system*, focusing on process modeling and automated execution of models for validation purposes, thus enabling direct process experience for all stakeholders.

The technical system was implemented prototypically using a novel type of workflow management system and then was tested to evaluate its impact on clinic operation. The tests were accompanied by an instrument for tailoring the technology being used and based on a formative assessment. The goal of the formative assessment was not only to improve the interactive handling of the tool support but also to monitor behavioral changes of the stakeholders involved in planning procedures when supported through the new system. Thereby, ongoing feedback could be provided that the development team used to further improve the quality of scheduling and that stakeholders used to deepen their understanding of the clinic's planning procedure.

2.2 Executing the Project

In the beginning of the analysis, the involved stakeholder (doctors, nurses, and administrators) needed to agree on the overall goal of the project, namely to optimize schedules for the sake of effective, high-quality patient treatment. Knowledge codification and semantic representation were performed using concept mapping, procedures used successfully by groups to develop a representation of a domain, situation, or workflow (Novak, 1998), and to capture content in its systemic context (Trochim, 1989).

Accordingly, when *capturing the state-of-operational-affairs in daily workforce planning*, the participants in the case study were asked to begin drawing by putting down node symbols using cards representing concepts or items meaningful for planning and identifying their mutual relationships on a virtual or paper surface, according to their experiential knowledge. They were also asked to draw the flow of communication between the stakeholders or technical systems with which they interact in the course of scheduling.

The mapping procedure was actually based on the Value Network Analysis proposed by Verna Allee (Allee, 2008). She distinguished between tangible and intangible deliverables that are exchanged among actors and represented as nodes in a dedicated concept map termed a holomap:

- *Goods, services, and revenue* are tangible exchanges between stakeholders for services or goods. These comprise all transactions involving contracts and invoices, return receipt of orders, and requests for proposals, confirmations, or payment. They also include knowledge products or services, once they generate revenue or are expected parts of services (e.g., reports) or when they are part of the flow of goods, services, and revenue.
- *Knowledge* captures all exchanges of strategic information, planning knowledge, process knowledge, technical know-how, collaborative design, policy development, and so on, which establish the flow supporting the core product and service value chain (Allee, 2008).
- *Intangible benefits* comprise a category of exchanges of value and benefits that go beyond the actual service, as they are not accounted for in traditional financial measures, such as the loyalty of customer or the sense for community (Allee, 2008).

A value network is composed of stakeholders, actors, or organizational units represented as nodes and relationships between them. They represent value exchanges of the types listed previously. Each of them is supported by some mechanism or medium, considered as enabler of a transaction to take place between actors. For instance, if two people want to exchange messages about a meeting, they may use the mechanism of e-mail or voice mail to support the exchange.

In the case study the offered coding options were a directed solid line ("tangible") for information they needed to deliver and receive according to the represented roles, whereas a directed dashed line ("intangible") for voluntary services or other noncontracted relationships. The following figure shows an administrator's representation. Typically, the person articulating work knowledge began by placing a symbol or node into the center of the map and labeling it with a person's name or some role designation, such as workforce planner.

Of utmost importance for systemic (i.e., contextual) development is to reveal essential relations and constructs, as these are substantial semiotic carriers. They should be labeled to convey meaningful information (Rentsch et al., 2010), such as passing on special patient requests between clinic administration and senior physicians. The participants in the case study mostly identified as nodes on their map roles and functional units involved in workforce scheduling. The relationships the participants set were either unlabeled, directed solid or dashed lines that conveyed implicit or explicit exchange of information, such as timeliness of getting informed or patient records. Relations of each kind allow the identification of communication patterns, which are central to coherent and consistent planning procedures.

Finally, the facilitators asked each stakeholder drawing an individual map whether their concept map represented all relevant elements of the organization from his or her perspective before proceeding with further analyses. In most cases, the participants then enriched their maps with additional auxiliary or enabling actors, such as stakeholders or technical systems with which they interacted in the course of workforce-scheduling processes. Notably, such maps correspond neither to traditional data modeling nor to business process modeling but rather combine structured elements from both, in particular the elements of roles and the mutually exchanged data between roles. As such, they provide a sound basis for detailing and completing fundamental aspects when (re)organizing work: organizational units, data, and the flow of control Fig. 6.1.

Once the created concept maps in our case study contained all relevant nodes (roles, actors, systems) and relationships, the communications behavior of each stakeholder became evident through the directed links between the nodes and could be further analyzed.

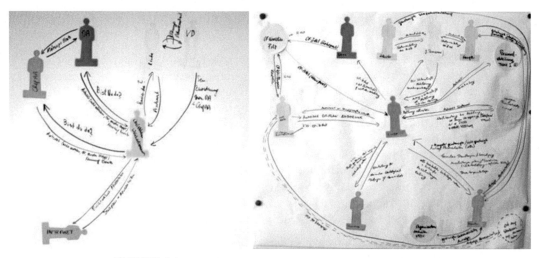

FIGURE 6.1   Sample concept maps—administrator perspectives.

A *consolidation of the state-of-affairs* in scheduling daily work revealed various sets of planning documents and procedures, as well as a lack of proper planning periods. In particular, the engagement of clinic stakeholders in academic teaching and research activities in the unit had not yet been taken into account in any of the planning procedures. When teaching academic courses, doctors were not available for patient service. This extension of the planning scope and period was identified as a requirement for future planning support. The corresponding planning systems of the adjunct university had not yet been coupled with the hospital's human-resource planning system.

A subsequent *communication analysis* was based on the input of seven stakeholders, with at least one member of each professional group asked to articulate their knowledge and perception of the procedure for daily workforce planning. Besides the causal links between actions (flow of planning), the outcomes were documented in terms of explicit or implicit deliverables (tangible and intangible exchanges between stakeholders). The reflection of the (re) presented meaning aims to justify the current organization of work tasks; otherwise, misinterpretations are likely to occur due to unreflected postprocessing of already generated knowledge (Sandberg, 2005).

According to the approach of Value Network Analysis, three steps need to be followed based on representation of the current state of work affairs in order to create the potential for change in a mindful way. The first analysis (exchange analysis) interprets the interactions of actors as perceived by individual stakeholders. Usually, patterns may be recognized of which stakeholders had not been aware. The subsequent analysis (impact analysis) reflects all received *inputs* from an individual stakeholder perspective. This is followed by the third analysis (value creation analysis), concerning the *output* of the reflected stakeholder interactions. Fig. 6.2 provides an overview of the sequence that was followed and each of the analyses.

Fig. 6.3 exemplifies the first step, the exchange analysis capturing formal, contracted, informal, or socially considered valuable deliverables among stakeholders (also termed

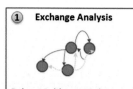

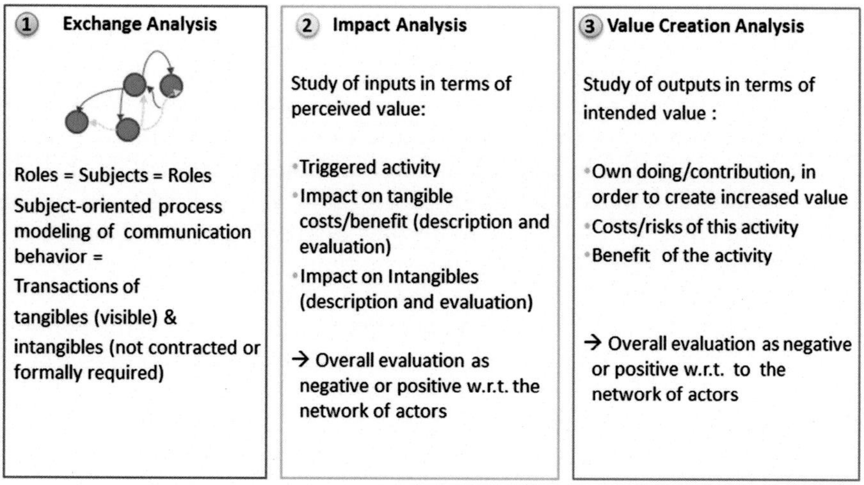

① Exchange Analysis

Roles = Subjects = Roles
Subject-oriented process
modeling of communication
behavior =
Transactions of
tangibles (visible) &
intangibles (not contracted or
formally required)

② Impact Analysis

Study of inputs in terms of
perceived value:

- Triggered activity
- Impact on tangible
costs/benefit (description and
evaluation)
- Impact on Intangibles
(description and evaluation)

→ Overall evaluation as
negative or positive w.r.t. the
network of actors

③ Value Creation Analysis

Study of outputs in terms of
intended value :

- Own doing/contribution, in
order to create increased value
- Costs/risks of this activity
- Benefit of the activity

→ Overall evaluation as negative
or positive w.r.t. to the
network of actors

FIGURE 6.2   Bridging the gap between existing situation and possible changes based on the specified actor network (concept map).

FROM	TO	Tangible	Intangible	Analysis
surgeon	administrator	request for support		lack of transparency w.r.t. availability
academic teacher	administrator		teaching duty	university schedule is not accessible
head of unit	administrator	contact request for medical expert	time pressure	disturbance of workflow
administrator	surgeon	status report		completeness of data set should be checked
administrator	head of unit	contact report or coordinates for contact	low confidence in results	quality of deliverables should be checked
administrator	HR department	request for scheduled duties	need of information	default transmission of scheduled duties for each person to administrator from HR department?

FIGURE 6.3   Sample exchange analysis sheet.

transactions among stakeholders). It reveals implicit relationships, denoted as intangibles that stakeholders consider relevant and that may be highly relevant for successful task completion on the organizational level (Nonaka and Von Krogh, 2009). The exchange analysis was structured according to the items represented as row headers in the table, capturing both parties involved in exchanges (sender and receiver), both types of transactions or exchanged deliverables (tangible and intangible), and the interpretation of the captured relationships, subsumed in this case as findings and lead time (before tasks can be accomplished). Significant findings that required further discussion and analysis were several informal exchanges that were facilitating the scheduling procedure and inefficiencies in terms of communication flow.

While both the exchange and the subsequent impact analysis refer to the planning procedure as-it-is, the value creation analysis allows the redesign of interaction patterns and, thus, of the organization of work. Each participant was thus asked to offer (additional) deliverables to other stakeholders in the network that he or she thought were of added value for others.

In our case, in the course of value creation analysis, many informal communication contacts were offered for conversion to formal ones. For instance, the schedule of academic teaching was offered as public information to all other stakeholders. Such a move corresponds to qualify intangible deliverables (there was no contracted obligation to make individual schedules public) as tangible.

By making a commitment to the offer, the individual stakeholder expresses willingness to change his or her interaction behavior. In this case, additional value was generated by extending the planning period with the academic year. Multiple offers for change were collected, which finally increased the transparency of the planning process itself.

However, not all involved stakeholders were willing to share what they were actually doing in the course of planning. One of the clinic heads who was responsible for assigning qualified personnel to daily operation insisted he continue to use his individual spreadsheet planning tool, even after recognizing that the data quality increased once the other medical experts provided their offers to share individual schedules. He took their offers but did not make one himself, even though several stakeholders had complained about exhaustive wait times before the head communicated decisions regarding change requests for scheduling.

As in most socio-technical system development projects, the organization of work must be negotiated mainly from a social perspective, rather than a technological one. In our case the project involved setting the stage for successful interactions and communication relationships between stakeholders rather than improving the functionality or usability of technical systems. However, since the value creation analysis is the trigger to start creating models that represent the workflow to be implemented, it needs later support from technical systems.

The *participatory design and prototyping of a technical system* was based on the collected commitments that needed to be negotiated before proceeding. In the case study, subject orientation was chosen, a modeling and processing approach that follows a communication-based flow of control (Fleischmann et al., 2012). In this approach the basic constituents of processes are subjects, which encapsulate behavior, either that of humans or provided by technical systems. Subjects exchange messages (deliverables, data) in the course of accomplishing work tasks.

This perspective is captured by the subject interaction diagram shown in Fig. 6.4, which contains all relevant actors and units (stakeholder roles) and technical systems represented

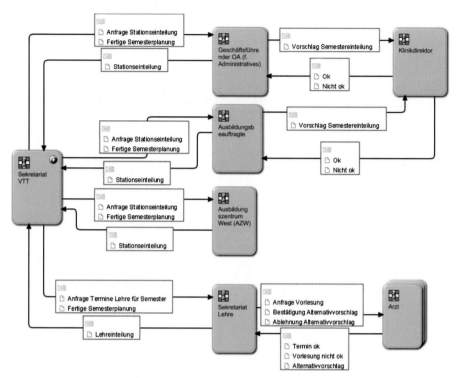

FIGURE 6.4   Communication requirements for daily workforce scheduling—subject interaction diagram.

as subjects (large rectangular symbols, such as Sekretariat VTT). These correspond to the nodes initially identified and labeled in the stakeholder concept maps. Subjects are related by message exchanges (small rectangles along directed links between subjects). The exchanges represent explicit interactions that have been initially specified in one of the concept maps or later in the course of the value creation analysis.

The subject interaction diagram in Fig. 6.4 captures all stakeholder roles relevant to workforce planning. The central role played by the clinic secretary (VTT) for collecting all schedule information and requests became evident in the course of value creation analysis. The relation of this role to other stakeholders also became more transparent, while the opaque role of the clinic head for planning remained.

For each subject (here, stakeholder role), a subject behavior interaction diagram must be specified, which details the encapsulated behavior as specified. Fig. 6.5 is an example subject behavior diagram for the task of arranging changes in the schedule. The diagram comprises three types of subject activity, namely (1) performing an action (doing); (2) communicating (sending or receiving a message); and (3) status information along the workflow (rectangles). In the case study, the specification process was facilitated by the project leaders and external partners.

When using subject-oriented business process management (Fleischmann and Stary, 2012), process models (subject interaction and behavior diagrams) can be executed without

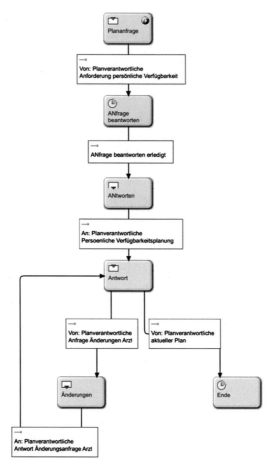

FIGURE 6.5   Sample encapsulated behavior specification—subject behavior diagram.

further transformation, once the set of message exchanges (i.e., send and receive operations) is complete. Stakeholders can be provided with hands-on experiences when running the processes that they have specified. In the workspace of a corresponding tool, such as Metasonic (www.metasonic.com), executable models become interactive elements that can be run at the click of a stakeholder.

Fig. 6.6 shows such an interactive design workspace. On the left side, the subject-oriented behavior model is shown, indicating the position of execution with a framed state. Its prototypical execution using a form management system, Metasonic Proof, is at the right side of the figure. Monitoring the execution of behavior diagrams triggers feedback loops and facilitates continuous codesign; modifications can be experienced immediately.

Prototyping a scheduling system based on processes also allows the design of schedules, such as the one displayed in Fig. 6.7. The prototype serves as a baseline for changes and provides stakeholders with a visual and experiential information object to handle along the scheduling process. Its embodiment in the communication-based scheduling workflow as

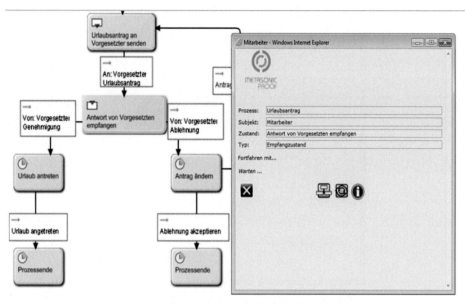

FIGURE 6.6   Hands-on experiences with business processes.

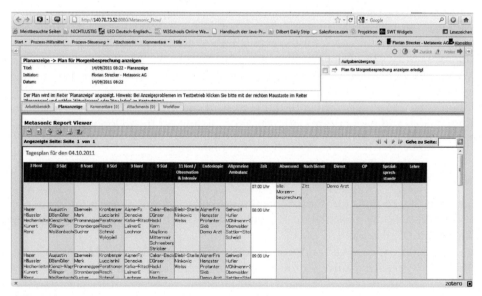

FIGURE 6.7   Workflow-based handling of schedules using Metasonic Suite (www.metasonic.de).

specified in the subject diagrams is shown in the upper part of the screen. At any time, users can recognize the state of execution, as the upper left panel conveys the task to be currently accomplished and the upper right panel contains all possible next steps that can be achieved by delivering the result of the work step to the next process actor.

FIGURE 6.8   Prototypical business object checking the availability of doctors in daily routine planning.

Exchanges between stakeholders are based on data encapsulated in business objects. For successful execution, the messages exchanged between subjects need to contain all relevant data, as exemplified in the check for the availability of doctors shown in Fig. 6.8.

In the case study, after implementing the prototype technical support as shown above, the stakeholders were asked whether the quality of planning could be increased through the redesigned processes. Of particular interest was whether the increase in transparency of communication could positively influence the perceived quality of planning. Each stakeholder, including doctors and patient care staff, was asked to answer the following questions:

- Do I communicate with all parties involved in planning?
- Are the generated schedules of use to me?
- What value added do I experience when using the schedules?

In addition, suggestions for improvement were collected with respect to data input (e.g., forms, search fields), the presentation of results and status information of processes, and the structure of communication.

The evaluation exposed usability problems of the workflow processing system; most stakeholders experienced difficulty accessing the technical system, although its behavior has been specified collaboratively. They also experienced difficulties handling the provided interactive features to accomplish their planning tasks. However, participants positively rated both the increase in scope with respect to planning, taking into account academic teaching periods, and the visibility and traceability of the work. The evaluation triggered further processes of redesign, both in terms of understanding the roles of actors—subjects needed to be refined—and the social adaptation to new behavior patterns or conventions.

In the case study, we used the subject-oriented workflow engine to generate prototypical workflow support, which then served as input for later project phases, including the formative assessment. After another year, the system was handed over to the IT department for embedding into the IT infrastructure of the hospital. The system still serves as a pilot for other clinics revisiting their planning procedures.

3. LEARNINGS FOR SOCIO-TECHNICAL DESIGN

Here, we reflect on the challenges with which the stakeholders were confronted and how they managed to overcome them. In particular, we discuss what kind of support stakeholders might need when becoming mindful designers of work processes, including their respective implementations using (novel) technical support systems. Since the models could be validated collaboratively and executed automatically after validation, the actual challenge in this case was to specify stakeholder and system behavior rather than implementing it on the technological layer.

3.1 Challenge 1: Dealing With Formal Positions (and Hierarchies)

Although hierarchies still seem to dominate in health care, when eliciting experience with existing work procedures and technical support systems, stakeholders' positions, especially for senior doctors, did not correlate to their depth of knowledge and willingness to share experiences in most settings. In our case, stakeholders strongly involved in daily patient care, and actually caring about patients, were highly active in the course of articulating existing work knowledge and developing novel solutions, regardless of their position in the formal hierarchy. The more insight into the implemented communication perspective a stakeholder had, the more active they were in listening and trying to grasp information reported by others.

The same holds true with respect to technology. Regardless of hierarchical position, staff closer to technology in operation were highly active during the design and redesign of process management support. These experiences indicate that socio-technical design processes need to take into account the reality that stakeholders are informed at different levels of competence and detail, according to the extent to which they perform tasks with immediate relevance for operating the clinic.

Consequently, support for mindful socio-technical design in planning requires, with respect to the (hierarchical) position of stakeholders:

- A common understanding of tasks and procedures on-the-fly must be assured. Whenever a model is created, either through articulation, representation, or processing, involved stakeholders may need to be briefed for the sake of common understanding, regardless of their hierarchical positions or qualifications.
- The setting should involve stakeholders from different hierarchical levels, units, and professions in order to recognize information and workflows beyond operative planning.
- In cases where hierarchical positions hinder the open exchange of experience and inputs for change, the facilitator must change the format to ensure either anonymity or equal opportunity to articulate and codesign.

- Cognitively demanding activities, such as following a certain planning logic, require means of visualization in the course of design, which keeps the logic available throughout the whole development process and serves as a baseline for collective design.
- Facilitators should point out that the value of articulated knowledge depends neither on the (hierarchical) position of stakeholders within the clinic nor on their professional and individual backgrounds.
- Knowledge must be elicited through an open format for articulation and collaborative reflection (semantic openness), as stakeholders should be able to start expressing themselves without long introductory statements; hospitals are under great time pressure.

However, for planning, doctors have the power to act not only according to their knowledge and experience but also according to their preferences, thus deviating from regulations or organizational particularities. Given such a context, mindful design aims to refer to situations where patients are affected by certain planning practices, for example, expert-based treatment for certain diagnoses. Still, facilitators need to accept process designs with a high degree of variability and nonrepetitiveness, such as ad hoc procedures, which seem to be inherent qualities of clinical planning processes.

3.2 Challenge 2: Ensuring Willingness to Use New Technology and Actively Participate in Evaluation

Even where health care workplaces are already equipped with different types of technology to support work processes, to introduce a new paradigm of thinking and subsequent way of acting requires substantial effort. In the case study, stakeholders were challenged to think in terms of (1) values rather than transactions and work results and (2) communication (thus, beyond conventional task accomplishment). The design of the technical system matched this idea of communicating rather than delivering.

As it turned out, employing new technologies that are not primarily focused on direct medical purposes, such as diagnosis and therapy, seems driven by traditional functional understanding of software, rather than workflow management, sequencing functions according to work procedures. Hence, using technology to support planning from a communication perspective seems to be perceived as disruptive or obtrusive by contrast to already recognized, well-established activity patterns, even if those patterns are not leading to satisfying (scheduling) results.

Several stakeholders continuously questioned the planning support tool and its underlying business logic. They argued it had cumbersome inputs and fewer control mechanisms compared to the previous planning procedure, beyond its usability issues. On one hand, we could observe that some stakeholders, in particular nurses, tried hard to embody thinking along the flow of communication. On the other hand, some stakeholders (workforce management) made no moves beyond their traditional understanding of the organization and structure of work processes. It required special effort to meaningfully embed novel design into their work.

In terms of refusal, stakeholders may find workarounds and placeholders to replace them in certain situations in order to avoid different styles of work and related technologies. In these cases, role models tend to play a crucial role, on both social and functional levels. In our case, the social level could be targeted through the clinic's peer group (e.g., opinion leader), and the functional level (e.g., medical experts) could be targeted through examples from other clinics that were undergoing similar processes. For those stakeholders who finally decided to participate in neither design nor implementation, for example, due to their position of power, the introduced methodological approach may not provide any means to change their behavior for the sake of higher quality.

Where people are reluctant to use a new type of technology, it can help to motivate them to recognize individual benefits and to start learning how to deal with the technology effectively. The same strategy can help if the new technology is not accepted at all. For all such cases, communities of practice should be offered, a format that facilitates building individual capacities in a self-organized setting.

3.3 Challenge 3: Effective Sharing—Articulation, Documentation, and Conveying Process Models Through Technological Artifacts

Documentation and communication are some of the key elements of (individual) articulation and sharing for collaborative reflection. All generated knowledge needs to be documented so that it can be used as a reference in reflection processes. In order to save effort and resources, the process of documentation should be as natural as possible once something to document occurs in the course of development. The current case suggests the following:

- Eliciting knowledge requires an open format for articulation and collaborative reflection (semantic openness). Hence, targeting process (re-)design through prespecified concepts and notations, for example, by using business process model and notation (www.bpmn.org), would restrict the articulation space and limit inputs for change. Moreover, the case study reveals that stakeholders initially consider functions (tasks) and organizational roles as integrated concepts, rather than separate model elements. Yet, sometimes it is necessary to delegate a task from one role to another.
- Knowledge codification must be accompanied by the sharing of knowledge. Knowledge must be accessible to others in order to enable collaborative reflection and codevelopment. Representations, such as concept maps or business process models, serve as baselines for discussion and discourse.
- Middle-out beats top-down and bottom-up analysis, reflecting social dynamics within the scope of modeling. Stakeholders begin modeling from their pragmatic perspectives and then challenge their interfaces, not only with respect to the received inputs but also by offering additional output to change the overall behavior of an organization.
- Intertwining the content or domain perspective with social processes not only helps to reflect a situation "as-it-is" in order to come up with ideas "as-it-could-be" but also helps to set the context of work procedures in terms of relevant factors for task accomplishment.

For effective participation in developing work (re)designs, stakeholders consider supportive means of reflection and creating opportunities. Of crucial importance seems to be role of the facilitator, who should precondition the process by clarifying the value of semantic openness when expressing experiences and ideas for change.

Another observation concerns the interface between individual learning and organizational development. Each mental model of a stakeholder requires place and space before starting the (re)design process. Even in cases with contrasting perspectives and model representations, the proposals interacting with change can be identical. In other words, stakeholders may begin from different points when reflecting the situation as-it-is while heading in the same direction when mindfully creating and publically committing to proposals for change.

After analysis of proposed modifications, semantically unambiguous process specifications that are stakeholder-conformed abstractions of reality can be expected and validated. Since these specifications are represented in terms of process models, they may need further transformation into practical experience. In cases with automated execution, collaborative effort among stakeholders can be supported by direct experience with the models as interactive work procedures.

4. CONCLUSION

Traditional approaches to process analysis and socio-technical system development tend to explain and document what is happening within an organization using formal modeling notation, such as the business process model and notation (www.bpmn.org). In order to avoid methodological bias and open up a value-based design space, concept maps and value networks allow focus step-by-step on existing interaction patterns as baselines for further development.

After analyzing exchange patterns and developing work output or outcomes with further stakeholders in a certain work context, organizations can decide which of these to promote for work process and artifact design. Process models, as stakeholder-conformed abstractions, should be automatically executable so that stakeholders can directly experience the redesigned work procedures. In complex settings, such as health care planning processes, such automatic execution could be the only way to judge whether the envisioned changes can become part of an organization's operation.

Several conceptual consequences can be drawn from the case study, according to the adopted perspective:

- For stakeholder-centered socio-technical design, organizations should be understood as social systems with communication as the prevailing activity for collaboration.
As communication systems, they can be considered to be completely closed systems comprising the components information, utterance, and understanding (including misunderstanding). As closed systems, they themselves specify their elements and structures (Luhmann, 1995). For that reason, organizations should be seen not as physical, autonomously existing and acting entities but rather as (communication) processes that are and must be continuously carried out in order to continue. Of utmost importance in that context is the recognition of relevant stakeholders, as they are the carriers of these communication processes. There may be relevant stakeholder even outside traditional organizational boundaries, such as patient in health care system, who may have direct behavioral impact, such as a patient canceling a clinical appointment.

- The smallest entity of all social systems is the communication act, which couples two or more actors or their communication acts. At its very heart is not the transfer of messages but the coordination of stakeholders (actors) in action. The operational processes that can be observed in an organization—meaning its patterns of action—can be explained as a result of communication. As Fritz B. Simon points out, communication is responsible for the dovetailing of actions of different actors, who are participants in communication (Simon, 2007).

Such a paradigm is compatible with existing approaches to the design of work procedures. For instance, business process models can reflect a communication perspective, either as choreography or orchestration of interactions between stakeholders (cf. www.bpmn.org). As Cohn and Hull (2009) demonstrated, (business) artifacts can be combined with data and processes as basic building blocks of modeling. Artifacts are key business entities (business-relevant objects) that evolve as they pass through a business's operation can be created, modified, and stored. Using this concept, business operations can be decomposed at various levels of abstraction.

Artifacts are typed using both an information model for data about the business objects during their lifetime and a lifecycle model that describes the possible ways in and times at which tasks can be invoked on these objects. In this case, the representation facilitates or strengthens the communication between business stakeholders in a way traditional approaches do not. As Cohn and Hull (2009) could show further, identified key artifacts are likely to become the basis of a stakeholder vocabulary. However, in this context, facilitation also seems to be critical, as the group must be formed into a team in order to create sustainable models (Hillier and Dunn-Jensen, 2012) while avoiding misinterpretations by external stakeholders in the course of their articulation (Sandberg, 2005). Further field tests should reveal effective means of intervention when guiding stakeholders to articulate their perceived realities of work.

References

Allee, V., 2008. Value network analysis and value conversion of tangible and intangible assets. Journal of Intellectual Capital 9 (1), 5–24.

Anyanwu, K., Sheth, A., Cardoso, J., Miller, J., Kochut, K., 2003. Healthcare enterprise process development and integration. Journal of Research and Practice in Information Technology 35 (2), 83–98.

Augl, M., Stary, C., 2015. Communication- and value-based organizational development at the university clinic for radiotherapy-radiation oncology. In: S-BPM in the Wild (35–53). Springer International, Berlin.

Bee, P., Price, O., Baker, J., Lovell, K., 2015. Systematic synthesis of barriers and facilitators to service user-led care planning. The British Journal of Psychiatry 207 (2), 104–114.

Cohn, D., Hull, R., 2009. Business artifacts: a data-centric approach to modeling business operations and processes. Bulletin of the IEEE Computer Society Technical Committee on Data Engineering 32 (3), 3–9.

Donetto, S., Pierri, P., Tsianakas, V., Robert, G., 2015. Experience-based co-design and healthcare improvement: realizing participatory design in the public sector. The Design Journal 18 (2), 227–248.

Fleischmann, A., Schmidt, W., Stary, Ch., Obermeier, St., Börger, E., 2012. Subject-Oriented Business Process Management. Springer International, Heidelberg.

Fleischmann, A., Stary, C., 2012. Whom to talk to? A stakeholder perspective on business process development. Universal Access in the Information Society 11, 125–150. http://dx.doi.org/10.1007/s10209-011-0236-x.

Gershengorn, H.B., Kocher, R., Factor, P., 2014. Management strategies to effect change in intensive care units: lessons from the world of business. Part I. Targeting quality improvement initiatives. Annals of the American Thoracic Society 11 (2), 264–269.

Ghazali, M., Ariffin, M., Amira, N., Omar, R., 2014. User centered design practices in healthcare: a systematic review. In: Proceedings of the 3rd International Conference on User Science and Engineering (i-USEr) (91–96). IEEE, New York.

Hillier, J., Dunn-Jensen, L.M., 2012. Groups meet … teams improve: building teams that learn. Journal of Management Education 37 (5), 704–733.

Lenz, R., Kuhn, K.A., 2004. Towards a continuous evolution and adaptation of information systems in healthcare. International Journal of Medical Informatics 73 (1), 75–89.

Lopes, M.A., Almeida, Á.S., Almada-Lobo, B., 2015. Handling healthcare workforce planning with care: where do we stand? Human Resources for Health 13 (1), 38.

Luhmann, N., 1995. Social Systems. Stanford University Press, Stanford, CA.

Nonaka, I., Von Krogh, G., 2009. Perspective—tacit knowledge and knowledge conversion: controversy and advancement in organizational knowledge creation theory. Organization Science 20 (3), 635–652.

Novak, J.D., 1998. Learning, Creating, and Using Knowledge: Concept Maps as Facilitative Tools in Schools and Corporations. Lawrence Erlbaum, London.

Prilla, M., Degeling, M., Herrmann, T., 2012. Collaborative reflection at work: supporting informal learning at a healthcare workplace. In: Proceedings of the 17th ACM International Conference on Supporting Group Work (55–64). ACM, New York.

Rebuge, Á., Ferreira, D.R., 2012. Business process analysis in healthcare environments: a methodology based on process mining. Information Systems 37 (2), 99–116.

Rentsch, J.R., Mello, A.L., Delise, L.A., 2010. Collaboration and meaning analysis process in intense problem solving teams. Theoretical Issues in Ergonomics Science 11 (4), 287–303.

Robert, G., 2013. Participatory action research: using experience-based co-design to improve the quality of healthcare services. In: Understanding and Using Experiences: Improving Patient Care. Oxford University Press, Oxford, pp. 138–149.

Sandberg, J., 2005. How do we justify knowledge produced within interpretive approaches? Organizational Research Methods 8 (1), 41–68.

Simon, F.B., 2007. Einführung in die systemische Organisationstheorie, vol. 1. Carl-Auer, Heidelberg.

Trebble, T.M., Heyworth, N., Clarke, N., Powell, T., Hockey, P.M., 2014. Managing hospital doctors and their practice: what can we learn about human resource management from non-healthcare organisations? BMC Health Services Research 14 (1), 566.

Trochim, W.M.K., 1989. An introduction to concept mapping for planning and evaluation. Evaluation and Program Planning 12 (1), 1–16.

Wong, E.L., Yam, C.H., Cheung, A.W., Leung, M.C., Chan, F.W., Wong, F.Y., Yeoh, E.K., 2011. Barriers to effective discharge planning: a qualitative study investigating the perspectives of frontline healthcare professionals. BMC Health Services Research 11 (1), 242.

Zhang, J., 2005. Human-centered computing in health information systems, part 1: analysis and design. Journal of Biomedical Informatics 38 (1), 1–3.

Dashboard Design for Improved Team Situation Awareness in Time-Critical Medical Work: Challenges and Lessons Learned

Aleksandra Sarcevic[1], Ivan Marsic[2], Randall S. Burd[3]

[1]Drexel University, Philadelphia, PA, United States; [2]Rutgers University, Piscataway, NJ, United States; [3]Children's National Medical Center, Washington, DC, United States

1. INTRODUCTION

Problems with productivity or safety in workplaces are often solved by introducing information technology (IT). These problems may arise due to inadequacy of existing artifacts or issues that have not yet been addressed by technology. In this chapter, we present our research on a clinical dashboard design that was motivated in part by observations about the use of existing artifacts and in part from the desire to provide new cognitive capabilities. Specifically we describe how we designed and evaluated TRU-Board, a dashboard system for improving team situation awareness during trauma resuscitation—a time-critical, high-risk, team-based, and information-intensive process of treating critically injured patients early after injury (Kusunoki et al., 2014). Our design approach was grounded in participatory design, allowing us to involve clinicians and domain experts in the system development, and to achieve common grounding across disciplines and among different stakeholders. Over 2 years, we engaged 49 clinicians from Children's National Medical Center in Washington, DC, who participated in a multiphase design and evaluation process, consisting of participatory design workshops, heuristic evaluation sessions, simulated resuscitations, video review of live resuscitations, and interviews.

The project built on several years of fieldwork that showed the need for an IT solution to address the challenges of information access and retention, team coordination, and team situation awareness (Sarcevic et al., 2008). A particular motivation was the observed lack of technology to support synthesis of patient and process information, as well as allow faster and easier contemporaneous access to information. To design effective computerized support, we needed to account for the larger system within which clinicians operate—often called a socio-technical system—and consider not only technology (e.g., software, hardware) but also people (e.g., clinicians, patients), work processes (e.g., workflow), organizational features (e.g., capacity, decisions about how health IT is applied, incentives), and the external environment (e.g., regulations, public opinion). For these reasons, we adopted a socio-technical perspective to designing the TRU-Board.

A *socio-technical perspective* considers the technical features of the system and social features of the work as fundamentally interrelated (Reddy et al., 2002), focusing not only on how technology will be developed but also how best to incorporate it into the environment and work activities. This approach has been used in other safety-critical work settings, including air traffic control (Bentley et al., 1992; Sommerville et al., 1993), underground traffic control (Heath and Luff, 1992), and financial systems (Hughes et al., 1994). Although adoption of this approach has been slow, researchers and practitioners in medical domains have also used it to design clinical systems (Berg et al., 1998; Berg, 1999; Reddy et al., 2002; Pratt et al., 2004).

The socio-technical challenges to designing systems in trauma resuscitation include the "messy" (Berg, 1999), ad hoc and unpredictable work processes with tight timelines, high risk of human error, and diversity of information needs for interdisciplinary teams. Hierarchical organization of trauma teams plays an important role as well, posing additional challenges to design. Other challenges include multiple information sources, different data types and modalities, and the need for the team to focus on the patient rather than on interaction with a computer system. Here, we describe how we addressed some of these challenges and the lessons learned through the design and evaluation process.

2. BACKGROUND: DOMAIN OVERVIEW AND APPROACHES TO DASHBOARD DESIGN

Trauma resuscitation is a fast-paced, high-risk clinical process requiring accurate and timely sharing of patient and process information for performing rapid diagnostic and therapeutic interventions. Unlike other clinical settings in which patient management relies on existing information, trauma resuscitation primarily relies on emerging information. The initial steps of trauma resuscitation focus on general assessment and management of major organs, such as ensuring an adequate airway, evaluating breathing (or respiratory) function, assessing blood circulation, and evaluating neurological status. This initial evaluation follows an ABCD algorithm (**A**irway, **B**reathing, **C**irculation, **D**isability) to quickly evaluate the patient's overall physiological state (American College of Surgeons, 2008). These steps are then followed by a secondary, head-to-toe assessment to identify other injuries and determine the plan of care. Resuscitations take place in the trauma bay, a designated room in the emergency department (ED), and usually last between 20 and 30 min. A typical trauma team consists of a team leader (senior surgical resident, surgical fellow, or an attending surgeon), an emergency medicine physician, a bedside physician (junior surgical resident), a respiratory therapist, an anesthesiologist, a technician, bedside nurses, and a nurse recorder (scribe).

The interdisciplinary, time-critical, and high-risk nature of trauma teamwork poses several unique challenges to display design. First, there is no consistent pattern of injury, making most resuscitation events unique and unpredictable (Fitzgerald et al., 2008). Second, clinicians are required to make critical decisions under high levels of uncertainty, often with minimal information and intense time pressure (Vinen, 2000). Information about the patient, such as their past medical history may be incomplete, misleading or simply missing. Consultants may not be readily available, often requiring decisions before the entire plan of care can be formulated. Third, trauma teams consist of personnel with different training backgrounds, experience levels, titles, and clinical disciplines. They must work together as a unit, yet team formation is ad hoc; as members are called from different units and departments, they may not have worked with each other before. Fourth, the trauma bay poses many physical constraints, with team members performing their tasks in close physical proximity around the patient and often competing for space and physical access to the patient, especially when treating younger children. Finally, clinicians are skilled, informed, and used to working in this type of environment, yet interventions may not be completed at the right time, in the right amount, or in the right order (Fitzgerald et al., 2008).

Prior work showed that resuscitation aids could provide an important advantage to the feedback loop by shortening decision implementation time and facilitating critical thinking (Luten et al., 2002). A dashboard displaying critical patient and process information during resuscitations that minimizes the team's mental effort to synthesize and retain the information holds promise as an approach to mitigating safety risks and improving outcomes.

2.1 Clinical Dashboards and Approaches to Design

Clinical dashboards are a category of health information systems that integrate information from multiple sources and display it in an easy-to-read, color-coded graphical form, providing relevant and timely information for decision-making (Dowding et al., 2014). Prior work showed that real-time surveillance dashboards complement existing decision support mechanisms, synthesize patient data for evaluation, and serve as an additional check to prevent propagation of errors (Waitman et al., 2011). Clinical dashboards have become increasingly popular as tools for communicating important patient and process information or performance metrics at a glance. To date, dashboards have been used to support clinical practice, research, and performance improvement in many medical settings, including emergency departments (Batley et al., 2011; Stone-Griffith et al., 2012), intensive care units (Egan, 2006; Starmer and Giuse, 2008; Salazar et al., 2011; Sebastian et al., 2012; Koch et al., 2013), and operating rooms (Levine et al., 2005; Meyer et al., 2005; Bardram and Nørskov, 2008).

Based on a socio-technical approach, designing clinical dashboards should, at a minimum, apply design principles and practices that are aligned with the way people see and think. Most studies of clinical dashboards, however, have focused on addressing the challenges in implementation (Batley et al., 2011), defining metrics (Stone-Griffith et al., 2012), or measuring post-implementation use effects (Starmer and Giuse, 2008), with only a few studies explicitly describing the design and evaluation process. Approaches to designing clinical dashboards also vary between studies. Some are promoting a socio-technical approach, recognizing that many current devices and systems for improving medical procedures are introduced in isolation (e.g., Egan, 2006; Craft et al., 2015). Others are accepting that dashboard designs often proceeded without experts in user interface design (e.g., Batley et al., 2011; Sebastian et al., 2012). Given their widespread use in patient-care settings, as well as an increased emphasis on systems safety (IOM, 2012), a recent review of clinical dashboards has called for researchers to establish clear guidelines for dashboard design (Dowding et al., 2014).

Most clinical dashboards have been developed for less dynamic environments than trauma resuscitation, primarily to help with data monitoring over longer time periods. Some, like those developed for EDs, also address the needs of providers caring for critically injured or ill patients, but their time scale is longer than that of trauma or emergency medical resuscitation. Drews and Diog (2014) developed a vital signs display for intensive care unit nurses. Their display organizes related variables in close proximity and integrates low-level data into more abstract representations, known as "configural display." Compared to vital signs data such as waveforms or time series that can be easily integrated into aggregate representations, a display like our TRU-Board shows mostly qualitative or categorical data about the patient evaluation or process status, which are harder to aggregate. Bardram and Nørskov (2008) developed a context-aware patient safety system (CAPSIS) for supporting surgical procedures in the operating room. Although CAPSIS is also intended for a safety-critical hospital setting, important differences in the nature of work exist between the operating and trauma rooms: surgical procedures are usually planned events with known information, whereas trauma resuscitations are unplanned with information being accrued in real time. In addition, CAPSIS is interactive and intended for use either before or while suspending interactions with the patient. In contrast, TRU-Board is intended for use during the actual patient interaction, without the need for team members to approach the display and interact with it. These domain differences resulted in different sets of requirements, with timeliness and unobtrusiveness playing a key role for our design.

Another challenging aspect of the TRU-Board design has been the discovery and selection of information items to display. Most research on clinical dashboards has considered how to visualize the information that is already in regular use, such as patient vital signs or demographic data. In contrast, we first needed to discover what types of derived patient or process information were most relevant for a space-constrained peripheral display. Unlike display design problems where the information items are known and the challenge is to identify the best visualization, our design problem required that we first determine what information items could be derived and then how best to display them.

3. TRU-BOARD DESIGN GOALS AND DISPLAY FEATURES

Medical providers participating in the design process perceived the overall purpose of the display as a tool to serve two main functions: (1) support awareness during the process and (2) improve team performance and communication (Kusunoki et al., 2015). The most common and agreed upon purpose of the display was to provide a quick overview of the resuscitation to make sure that everyone was on the same page. Specifically the display was intended to serve as a reminder of what had been done, what was in progress, and what had been ordered or was pending. Participants also felt that the display would help facilitate communication and provide a mechanism for double-checking teamwork by making verbal information more persistent. Based on these perceptions, two major design goals emerged for the display: (1) providing basic overview of the resuscitation progress and facilitating

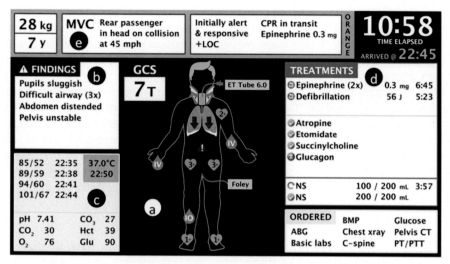

FIGURE 7.1 Final TRU-Board display design and its components: (a) patient body graphic with major findings and treatments; (b) list of major findings; (c) patient data trends and vitals; (d) list of treatments; (e) patient demographics and prehospital data.

periodic process summaries and (2) reducing redundant communication. We next describe the components and features of the TRU-Board display and how they address these design goals (Fig. 7.1).

3.1 Providing Basic Overview of the Resuscitation Progress and Facilitating Periodic Process Summaries

A common practice for trauma team leaders is to provide a process summary (and sometimes multiple summaries) at the beginning, middle, or end of the resuscitation, listing major findings, critical vital signs, treatments and interventions, tasks in progress, and incomplete orders. Teams can then take a brief step back and revisit the "big picture." We designed three display elements to address this goal (Kusunoki et al., 2014). First, completed tasks and steps are conceptualized through a visualization of the patient body with major findings, tubes, lines, and drains depicted (Fig. 7.1a). Major findings are also listed separately on the left side (Fig. 7.1b), and trends in vitals are listed in "findings" (Fig. 7.1c). Administered medications and fluids are combined into one running list called "treatments" on the right side of the display (Fig. 7.1d).

3.2 Reducing Redundant Communication

Critical prehospital information about the patient is reported at the beginning of the resuscitation, as the Emergency Medical Services team hands the patient over to the resuscitation team. Patient information includes demographics (e.g., age, weight), mechanism of injury (i.e., how the patient got injured), prehospital interventions, and en-route changes in patient status. Our participants emphasized the importance of including this information on the display for two reasons. First, because patient information is reported early in the event and only once, team members have difficulty accessing these data later as they evaluate and treat the patient. For example, bedside nurses requested displays of age and weight to reduce the need for questions about these parameters when they drew medications or prepared fluids (medication dosages and fluid volumes depend on the patient's age and weight). Second, ad hoc–team formation makes it common for some team members to arrive later than others and miss important information (Lee et al., 2012). When team members arrive late, the team leader must temporarily shift focus to update latecomers about the patient's status.

We implemented two display features to reduce redundant communication. First, at the top of the display header, we showed patient information such as age, weight, mechanism of injury, prehospital interventions, medical history, timer, and arrival time (Fig. 7.1e). Second, using an image of the body with visual representations of abnormal findings and procedures, we provided a snapshot of the patient status (Fig. 7.1a).

4. TRU-BOARD DESIGN AND EVALUATION PROCESS

Over the course of 20 months (November 2012–July 2014), we conducted 11 design and evaluation phases, consisting of participatory design workshops, simulation sessions in the trauma room, heuristic evaluation sessions with interviews and video review, and a focus group (Fig. 7.2). The study protocol was reviewed and approved by the hospital's Institutional Review Board (IRB). Below we describe this process in greater detail, including the research site, participants, and specific methods we used throughout the study.

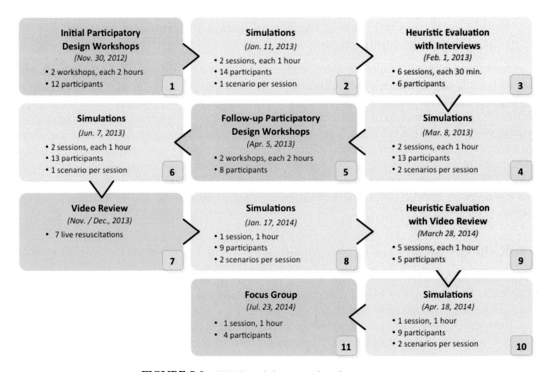

FIGURE 7.2 TRU-Board design and evaluation process.

4.1 Research Setting

We conducted our study at Children's National Medical Center, a freestanding pediatric hospital with a level I trauma center in Washington, DC. Level I is the highest designation for a trauma center that provides definitive, comprehensive care for every aspect of injury (American College of Surgeons, 2006). Each year, Children's National Medical Center admits over 1000 children with trauma and burn injuries. Injured children initially triaged as having suspected or observed severe injury are first treated in a resuscitation room in the emergency department by a multidisciplinary team.

The resuscitation rooms contain several IT systems and tools that assist teamwork during resuscitations. These systems include vital signs monitors and large wall displays to augment the view of the vital signs or the patient. The displays are positioned at the front and back of the rooms, allowing each team member to have an unobstructed view of the displayed information regardless of their location in the room. Because time is a critical factor in high-risk, safety-critical patient management, teams rely on several temporal artifacts, including clocks showing the absolute time, stopclocks showing the resuscitation time (i.e., time since the resuscitation started), and timers counting down from a specified amount of time (e.g., automatic readout of blood pressure every 5 min). Several additional handwritten and paper-based tools are available for obtaining and displaying information, including a dry erase board for displaying the patient's weight, a trauma flowsheet for manually documenting the process, paper-based wall charts for looking up information on treatment parameters, and a paper-based trauma resuscitation checklist for ensuring compliance with the resuscitation protocol.

4.2 Participants

Forty-nine participants were recruited to take part in different design and evaluation phases throughout the study (Table 7.1). Each phase involved the core team roles required

TABLE 7.1   Participant Demographics

Roles	Participants (N = 49)	Average Experience (Years)
Anesthesiologist	6	5 (1/6 not reported)
Bedside nurse	7	9 (3/7 not reported)
Emergency medicine physician	6	4.5 (1/6 not reported)
Physician surveyor	8	3 (4/8 not reported)
Respiratory therapist	6	4 (1/6 not reported)
Scribe nurse	7	17 (2/7 not reported)
Surgical team leader	9	2 (3/9 not reported)

during trauma resuscitation: anesthesiologist, bedside nurses, emergency medicine physician, physician surveyor (nurse practitioner or surgical resident), respiratory therapist, scribe nurse, and team leader (surgical fellow or attending). All physicians and nurses who regularly perform these roles at our research site were eligible for participation.

Because recruitment is challenging in this environment, we allowed repeated participation between but not within phases, with 22 participants attending more than one design or evaluation session throughout the study. Because participant recruitment was most challenging for simulated resuscitations, three hospital-based members on our research team acted as confederates, filling in for medication nurse, scribe, and second bedside nurse roles in four out of five simulation sessions.

4.3 Methods

We next describe the participatory design approaches we used to elicit and evaluate design ideas, information needs, critical information items to display, and their layout. The most common setting for participatory design is the participatory workshop. These ideation sessions were followed by rapid prototyping and formative evaluation using simulated resuscitations and heuristic evaluation combined with interviews and video narration (Fig. 7.2). Further design iterations were performed after each evaluation session, yielding 16 major design iterations throughout the study and multiple minor iterations in-between. Although primarily designed as formative evaluation sessions, simulated resuscitations, interviews, and video narration were also used to further elicit information needs, validate findings from previous sessions, gain insight into team awareness needs, and assess the feasibility of the clinical dashboard for trauma resuscitation.

4.3.1 *Participatory Design Workshops*

The design process started with two participatory design workshops, each involving a group of core trauma team roles. The goals were to elicit perceptions about the resuscitation process and what information was critical to the work, ideas for dashboard design, and any concerns about using a clinical dashboard during resuscitations. To provide an environment where users with diverse perspectives have equal opportunity to engage in the design process, we used an approach to participatory design called PICTIVE (Plastic Interface for Collaborative Technology Initiatives through Video Exploration) (Muller, 1993). Our participants used low-tech design objects such as colored pens, pencils, scissors, pieces of paper, and Post-it notes to create dashboard design sketches and prototypes. Discussions during the workshops, especially those surrounding the design, were audiotaped and videotaped to provide an informal design rationale when reconstructing the purpose of specific design elements or choices. Design outputs such as individual sketches and group designs were photographed to preserve the layout of design elements and facilitate our analyses of role- and team-based information needs.

Each workshop lasted 2 h and was split into five different activities that built on each other: quick survey, individual design session, group design session, information ranking, and discussion of concerns. The main purpose of the survey was to help participants ground their design thinking based on actual scenarios as they reflected on their most recent resuscitations.

Following the survey, participants were given sheets of construction paper to think about the pieces of information that are critical to their work and sketch a design for their personal information dashboard. This strategy allowed us to understand the detailed role-based information needs that may be lost through group design activities. Participants then discussed individual designs with the group, and their sketches were posted on the wall for reference during subsequent workshop activities. After a short break, participants worked together as a group to create a dashboard design that incorporated ideas from their individual designs. This activity prompted participants to discuss their decisions and reach consensus on the most important design elements and features that would address the main information needs of all roles. We used group designs to better understand the information needs shared among roles. Once the group design was completed, we asked participants to rank the information pieces on their group display based on perceived importance of each to their role. The ranking activity allowed participants to voice an opinion about the most critical information pieces, despite any differences in power and outspokenness. We used the rankings to prioritize design features and information types for inclusion on the dashboard. The final activity involved a group discussion of concerns and any issues that participants anticipated with the dashboard use during real-world resuscitations.

Two follow-up design workshops were conducted mid-way through the design process, after several design iterations and evaluation sessions (Fig. 7.2). The overall format was similar to the initial workshops but focused on eliciting input on display functionality. The first half of the workshop was used for "member checking"—a review and discussion of what was learned and accomplished, design ideas that emerged, and critical pieces of information that were included on current designs. We then engaged participants in the group design session to understand dashboard functionalities, dynamics of information, and information presentation. We discussed several issues, including what information is displayed at all times and what information changes, when information changes, how it changes, and how information is presented. To conclude the design session, we asked participants about the most important functions to their role and how would these functions be useful to their work. The workshop concluded with participants' feedback on issues and concerns about the dashboard's potential use in a real-world setting.

4.3.2 Simulated Resuscitation in the Trauma Room

High-fidelity simulations were used to evaluate the usefulness and perceptibility of our display system in realistic and challenging clinical scenarios. Throughout the design process, we performed five sets of simulations (phases 2, 4, 6, 8, and 10 in Fig. 7.2), with each set occurring after we rapidly prototyped and implemented designs from ideation or heuristic evaluation sessions. Simulation sessions were conducted in the main resuscitation room of the hospital's emergency department with the equipment normally available to teams. We used the wall monitors on both sides of the room to display our prototypes.

Each hour-long simulation session involved an entire resuscitation team participating in several activities. We first provided an overview of the research, session activities, and patient mannequin. Following this brief overview, we oriented teams to the display system, demonstrating its features and functionalities. Participants then performed two to four simulated resuscitations for about 15 min based on clinical scenarios ranging from moderate to demanding that were developed by medical experts on our research team. After the simulations, participants filled out a survey, providing individual feedback about their experiences using the display. A group discussion followed, asking for perceptions about (1) team communication and performance, (2) information on the display, and (3) display design and functionality. Participants then placed color-coded stickers based on their role to indicate up to five display features they found useful for their work and up to five features they did not find useful. If an information feature did not receive a vote, we considered it neutral.

Because our goal was to experiment with different data capture mechanisms for the entire system development, data capture varied between simulation sets. During the first set of simulations, we inputted information onto the display using a digital pen and paper flow-sheet used by scribes. During the second set, the prototypes captured data from digital pens from both the flowsheet and the leader's paper checklist. In both sets, scribe nurses and leaders were instructed to write on the flowsheet and checklist as they would normally do. To capture handwritten information and unstructured data such as progress notes, administered medications or fluids, we used the Wizard of Oz approach by having one researcher entering information via a dummy sheet that contained a select set of fields from the trauma flowsheet designed specifically for digital pen capture. Although feasible for capturing some information types (e.g., check-marked or circled information), the digital pen technology had several limitations, including inaccurately captured handwritten information, stray marks that triggered displaying wrong information, and inability to update information once it was recorded on the flowsheet. To address these challenges, we designed and developed a computer interface for entering the display information. This interface was essentially a computerized flowsheet but contained only data entry points that corresponded to the display information. Using the Wizard of Oz approach, a physician researcher was inputting display information during our third, fourth, and fifth simulation sets.

4.3.3 Video Review of Live Resuscitation Events

We used video review of live resuscitations to gain a better understanding of the trauma resuscitation process, work challenges, and the nature of awareness in real-life situations. We transcribed videos from seven events that occurred over a 1-month period. Due to patient privacy and IRB restrictions, videos of live events could be reviewed on-site only, requiring detailed transcriptions of all events, including all dialogue and activities. Field notes about the nature of the event and prominent initial findings were also recorded to help us understand the overall context for each event.

4.3.4 Video Review Sessions With Interviews

We conducted video review sessions with five trauma team members to (1) understand the perspectives of clinicians about maintaining awareness during the resuscitation process, (2) verify our observations and conclusions from previous sessions, and (3) elicit feedback on specific features of the most recent version of the display design. Each participant first narrated a 10-min video of a simulated resuscitation performed during simulation session 3, pausing the video when appropriate and commenting on aspects of the resuscitation that were unusual or important. We concluded by asking clinicians to perform a heuristic evaluation of the most recent display design.

4.3.5 *Concluding Focus Group*

We completed the display design cycle by conducting a focus group with four clinicians—an emergency medicine physician, a surgical coordinator, a bedside nurse, and a scribe nurse. The purpose of this focus group was to elicit feedback on the latest design iteration and finalize the display design. We started the session with a demonstration of the display functionalities using a clinical scenario. After the demonstration, participants marked the display features they found useful and not useful on a paper-based version of the display using color-coded stickers based on their roles. We also asked participants to review the body icons for clarity. The session concluded with a discussion about the participants' concerns about using this display and an invitation to provide their thoughts on possible future directions of the project.

4.3.6 *Data Analysis*

Analyses were conducted to identify trends on all artifacts, design prototypes feedback, and other data collected throughout the process. Individual designs from workshops were transcribed into a matrix to analyze the information features each role included in their design. The features were then grouped by type and sorted by number of times that they were included. Group designs were transcribed in a similar manner by grouping information features by type and recording the top five ranks that each role assigned to features. Usefulness ratings assigned to information features during simulations were also calculated to analyze how perceptions of the display changed over time (i.e., after major design iterations). Discussions were transcribed and analyzed using an open-coding technique to identify salient themes and to supplement the findings from design artifacts, feedback, and prototypes.

4.4 Summary of the Design Process and Outcomes

The hierarchical nature of trauma teams and the multiplicity of responsibilities, disciplines, and training levels naturally lead to a diversity of information needs. Patient data that are meaningful to one team member might go unnoticed by other team members. While each role has particular information needs, we also observed several overlapping needs among roles that need to be met to coordinate tasks. This mix of information needs also became evident as we were designing and evaluating display prototypes, revealing both role and design tensions (Kusunoki et al., 2014). We identified eight categories of information based on participants' group designs during the initial participatory design workshops: (1) patient demographics and prehospital information; (2) vital sign values, waveforms, and trends; (3) findings from the primary survey (i.e., ABCDE steps); (4) medication names, dosages, and administration times; (5) procedures: types and locations of tubes, lines, and drains; (6) laboratory and radiology orders and results; (7) fluid types and amounts; and (8) disposition plan.

Although it is difficult to reconcile various information needs and mitigate role hierarchy when developing a shared display, we applied two strategies to minimize the effects of these factors. First, each participant created their ideal information display to suit their role, discussed the various information features suggested in individual designs to reach consensus, and then created a design as a group. This strategy allowed us to understand the role-based information needs that may be lost through group design. Second, we encouraged participants to include as many information features as possible when designing a group design to support all roles because they would be able to individually rank the top five information features. This strategy minimized the influence of hierarchy and outspokenness while identifying individual priorities and those shared across roles by comparing rankings on group designs. We used a similar strategy in simulation sessions. Instead of ranking their top information items, participants used checkmarks and crosses stickers to indicate information features they found useful or not useful. To quantify the effect of these approaches, we analyzed the individual designs and compared them to the consensus-based group designs to determine whether some roles compromised more than others, that is, if fewer information items suggested in their individual designs propagated to the final prototype design. Rankings from simulation sessions also allowed us to see whether the roles that compromised the most during the group design were also the least satisfied with the final display design. Results from these analyses showed that our final display design included most of the information features from individual designs, and each role compromised on only three features or less (Kusunoki et al., 2014).

5. SOCIO-TECHNICAL CHALLENGES IN DESIGNING DASHBOARDS FOR SAFETY-CRITICAL MEDICAL WORK

Throughout the design process, we evaluated the display design several times but mostly focused on the formative evaluation, ensuring that features and functionalities were meeting the user needs and were easily perceivable and understandable. In short, we focused on "getting the right design" (Tohidi et al., 2006), presenting multiple design alternatives to elicit critical feedback from users and to determine the elements of the right design. Because of technical challenges of automatically capturing information for the display, we have postponed our summative evaluation of a fully functional system to analyze the display effects on team performance. As we discussed our potential approaches to summative evaluation, we found that selecting appropriate metrics from literature or identifying new metrics was not straightforward. Planning and designing the study was also challenging because multiple approaches can be used to evaluate complex socio-technical systems. The impact of our display design is difficult to measure or quantify because of the complexity of the environment, variability in patient cases and attributes, as well as variability in provider expertise and experience. We also encountered providers' resistance toward real-world deployment because the display did not undergo safety checks and we had no evidence that the display was safe to use in actual resuscitations.

Below we describe these two socio-technical challenges that emerged in our design: (1) measuring display effects on team performance, where "socio" is about the system effects on practice and (2) real-world deployment and adoption issues.

5.1 Challenges in Measuring Display Effects on Team Performance

Evaluating the effects of a dashboard-type wall display on trauma team performance requires assessing how well the display solves the perceived problems with performance and safety that arise with the existing technology or its absence. We need to show that providers can extract clinically useful information from the dashboard and apply it in their work. We therefore considered metrics that measure whether the dashboard substitutes the existing artifacts and information sources in the trauma room or complements these existing sources with new capabilities. The main information artifacts in the trauma room are currently either paper-based (e.g., trauma flowsheet and checklist) or low-tech (e.g., wall clocks and a whiteboard with prehospital data about the patient). In addition, trauma teams rely on collective memory and verbal communication to exchange information (Sarcevic et al., 2008).

To measure the extent to which our dashboard substitutes existing artifacts and technologies, we need to assess whether the amount and patterns of interaction with the existing artifacts change as a result of the dashboard introduction. We can also expect that the dashboard may change communication patterns and reliance on collective memory. In contrast, to measure the extent to which our dashboard complements existing artifacts, we need to consider providers' interaction with the dashboard and their perceptions of its usefulness. Metrics that we considered include the following.

Quantifying flowsheet lookups: We observed that team members, and especially the team leader, reference the paper flowsheet to access the information they forgot or missed. This referencing may be direct (i.e., by approaching the recorder's desk and looking up) or indirect (i.e., by inquiring the recorder) (Sarcevic and Ferraro, 2017). A possible metric for display effects on decision-making would then be the number of times the leader references the flowsheet compared to the number of references to the flowsheet after the needed information was displayed. This metric can be implemented by reviewing videos of resuscitations and marking leaders' references to the flowsheet with and without the system. A potential drawback of this approach is that flowsheet lookups focus on the leader, so we need to supplement it with other metrics to determine the value of the display for the team.

Quantifying checklist notes in the margins: We observed that team leaders frequently make notes on the paper-based checklist that is now mandatory during resuscitations in the hospital where this research was performed (Sarcevic et al., 2016). Given this observation, another possible metric for display effects would then quantify the amount of notes on the checklist when the display is present. A positive effect of display would be a finding that shows fewer notes on the checklist when the display is available because the display now serves as team leaders' external memory instead of the checklist. Content analysis of these notes could help determine if the amount of specific note types declines or disappears after the display is deployed. This qualitative analysis could also help with the iterative design to identify display items that need improvement to cause observable effects on team leader's work. A drawback of this metric is that it also focuses on the team leader.

Quantifying whiteboard use: Patient weight is particularly important for pediatric patients when determining the correct dosage of medications. When known, the weight is usually written on a whiteboard located in the trauma room. It is also part of our current display design (Fig. 7.1). A potential metric for display effects would quantify the whiteboard use before and after the display deployment. This metric would determine if the whiteboard use and inquiries about the patient's weight decline or disappear after the display is deployed, indicating again that the display took over the external memory role from the whiteboard.

Quantifying the number of inquiries: We could also quantify questions that team members ask about the information items available on the display before and after the display is deployed. Here, we could examine inquiries supporting two types of awareness:

- Inquires about task awareness or the status of tasks, especially precondition tasks that need to be completed before another task can be performed. Example inquiries include "Do we have intravenous access yet?" or "Are the medications in yet?"
- Inquiries about patient data awareness or information such as the mechanism of injury, prehospital information, or allergies. These could also include questions about a specific type of information that had been reported earlier by another team member. Example inquiries include "What's the mechanism?" or "Do we know of any allergies?"

Display looks: Measures of team members' glances at the display would show how the frequency and duration of glances change over the course of the deployment period. For example, we can assess whether teams look more frequently or for longer periods of time or determine who in the team is looking at the display and when in the process? We could implement this metric by using portable eye trackers to obtain precise measurements of when and for how long each team member made eye contact with the display. A less costly, though, more cumbersome approach would be by using video review (Kusunoki et al., 2013).

User perceptions: While less objective than the above metrics, user perceptions can provide significant insight into the display effects. We have considered several opportunities to unobtrusively obtain user feedback while the display is being deployed:

- *Poster annotations*: As a trial, we printed a large size poster showing the display and posted it in the lounge where emergency department staff spend time during breaks. We invited providers to annotate the poster with comments about aspects of the display they found useful or not useful to their work. Over the course of several weeks, we only received a few comments, mostly being very general such as "this is cool," or "I don't think this will work." It would be important to identify better ways to encourage and incentivize providers to provide constructive feedback using this method.
- *Clinical video review sessions*: Clinical video review sessions occur regularly at our hospital and provide an opportunity for providers to discuss their performance in an informal manner. User feedback could be obtained by observing these sessions to see if clinicians discuss the display and then asking questions about the display toward the end of the session.
- *Simulations and training sessions*: Because the display will be deployed in the real world, regular simulation and training sessions will also need to include the display. This mechanism could also be leveraged to include several display-related questions during the post-session debriefings.

The metrics we discussed above are necessarily narrow in scope. We can only measure low-level effects (e.g., number of lookups or notes on the checklist, or verbal communications) to indirectly infer if the display is supporting team awareness and improving patient outcomes. It will be challenging to know the extent to which display actually does improve patient outcomes and by how much. Similar challenges have also been observed in prior research on evaluating the effectiveness of medical systems such as electronic health records and checklists (Kramer and Drews, 2016).

5.2 Challenges to the Real-World Display Deployment and User Adoption

Data capture remains a key technical barrier to the real-world deployment of our display system. We initially experimented with digital pen technology as a method of capturing information written on the trauma flowsheet. Early in the design process, however, we found that capturing information with a digital pen directly from the paper flowsheet as the scribe was writing it was not feasible (Sarcevic et al., 2012). The digital pen limited the kinds of information we could accurately capture and display to a select set of flowsheet fields, which did not fully match the information types needed by teams. We also tried a dummy sheet designed specifically for digital pen capture, but this solution required an extra person to enter the information. We decided to instead use the Wizard of Oz approach to obtain the information for the display in a timely manner during evaluation sessions. In other words, we decided to decouple the information acquisition problem from the display design problem so that we could first get the design right and address the capturing issue later.

As we entered later stages of the design process, we revisited the challenge of data capture. One of the issues was that we did not have the resources to develop a tool for the scribe that could simultaneously serve the documentation and display purposes. After all, the goal of our project was not to develop a "documentation tool" because it effectively meant designing another user interface for the scribe, which represented a significant task. Although our initial efforts focused on developing a simple tool to only obtain information that our display showed, the real-world deployment would require more resources and more people involved for this interface to meet all requirements. First, we would need to integrate information entered by the scribe into the tool with both the display and the medical record. Second, all the fields currently on the flowsheet would need to be programmed into the input interface, rather than just a select set of information items needed for the display. The complexity of this task showed that keeping the scribe nurse as the main proxy for capturing the display information eventually led to the development of an electronic trauma flowsheet, which was a whole new research aim. For this reason, we decided to keep the scribe out of the loop and instead introduced a new role through the Wizard of Oz approach.

The Wizard of Oz approach also posed several challenges. First, we needed to decide whom to assign this role. Both emergency medicine physician and surgical leader are busy with delegating tasks and making decisions. In addition, the surgical leader is using the paper-based checklist. The scribes also have their own documentation work. Other team members are hands on during patient care. While busy team roles cannot engage in direct information capture or input, new technologies like speech- or gesture-based input are creating opportunities for their involvement in this process (e.g., O'Hara et al., 2014).

Although the Wizard of Oz approach solved (though only temporarily) the challenge of data entry, it did not address the "socio" challenges to the real-world system deployment. One adoption barrier was the lack of evidence that providers knew how to safely use the system—to know exactly where to look for specific information types, to correctly interpret meanings of all symbols and icons, and to trust the accuracy of displayed information. Because deploying the entire system in the actual work setting is challenging or even infeasible, we may need to deploy the system incrementally, to gradually introduce display components with incrementally added capabilities. This approach is also complex because we do not know in which order to add the capabilities, and we do not know if adding more capabilities will produce a linear effect on providers' work. In addition, as previously observed, isolating specific effects in an in-the-wild study is difficult. Rather, researchers have to make sense of data in the wild, with many factors and interdependencies at play causing the observed effects (Rogers, 2011).

Another adoption barrier was the lack of evidence that the display did not obstruct acquiring information that was not displayed. In other words, we needed to show that teams could pick up on the information that was not visible through the display but was critical to patient care. We discussed several ways to measure successful acquisition of nondisplayed information: (1) analyzing communication and monitor looks to determine how providers talk about the information that is not displayed but needed and how they behave when they cannot find the information on the display; (2) using the Situation Awareness Global Assessment Technique (SAGAT), an adapted human factors tool that determines team member knowledge of information about the patient scenario with and without the display.

We trialed SAGAT during the formative display evaluation in two simulation sessions to determine how the information display affected situation awareness (Endsley, 1995). At critical points in the resuscitation, we paused the team and blanked out the display. Each team member was then given a packet of paper on which to record his or her answers to questions about information that emerged during the resuscitation included on the display (e.g., "What is the current glasgow coma score of this patient and which medications and fluids have just been administered?"). This activity allowed us to determine that the technique would be useful for future implementation during the summative evaluation phase of the project.

6. CONCLUSION AND FUTURE WORK

Using the socio-technical perspective and through participatory design workshops, heuristic evaluation, and simulated resuscitations, we designed a clinical dashboard for supporting complex teamwork during medical emergencies. Taking an iterative participatory design approach was critical to arriving at a balanced design that considered role-based differences in information needs and team hierarchy. Over the course of the project, we faced two major socio-technical challenges in designing and evaluating the display: (1) measuring display effects on team performance and (2) real-world deployment and adoption issues. The next phase of our research will focus on implementing the dashboard prototype for use during real trauma resuscitations.

Acknowledgments

This research was supported by the National Library of Medicine of the National Institutes of Health under Award Number R21LM011320-01A1 and by the National Science Foundation Award No. 1253285. The authors thank medical providers who took part in this study. We also thank our collaborators and the staff of the Division of Trauma and Burns at Children's National Medical Center for their assistance in this project. Thanks also to Diana Kusunoki, a PhD student at Drexel University at the time, who contributed to this project as part of her PhD dissertation work.

References

American College of Surgeons, 2008. Advanced Trauma Life Support® (ATLS®), eighth ed. American College of Surgeons, Chicago, IL.

American College of Surgeons, 2006. Resources for Optimal Care of the Injured Patient. American College of Surgeons, Chicago, IL.

Bardram, J.E., Nørskov, N., 2008. A context-aware patient safety system for the operating room. In: McCarthy, J., Scott, J., Woo, W. (Eds.), UbiComp '08. Proceedings of the 10th International Conference on Ubiquitous Computing, Seoul, Korea, 21–24 September, 2008. ACM Press, New York, pp. 272–281.

Batley, N.J., Osman, H.O., Kazzi, A.A., Musallam, K.M., December 2011. Implementation of an emergency department computer system: design features that users value. The Journal of Emergency Medicine 41 (6), 693–700.

Bentley, R., Hughes, J.A., Randall, D., Rodden, T., Sawyer, P., Shapiro, D., Sommerville, I., 1992. Ethnographically-informed systems design for air traffic control. In: Mantel, M., Baecker, R. (Eds.), CSCW '92: Proceedings of the 1992 ACM Conference on Computer-Supported Cooperative Work, Toronto, Ontario, Canada, 1 – 4 November, 1992. ACM Press, New York, pp. 123–129.

Berg, M., August 1999. Patient care information systems and health care work: a sociotechnical approach. International Journal of Medical Informatics, 55 (2), 87–101.

Berg, M., Langenberg, C., Berg, I., Kwakkernaat, J., October 1998. Considerations for sociotechnical design: experiences with an electronic patient record in a clinical context. International Journal of Medical Informatics 52 (1–3), 243–251.

Craft, M., Dobrenz, B., Dornbush, E., Hunter, M., Morris, J., Stone, M., Barnes, L.E., 2015. An assessment of visualization tools for patient monitoring and medical decision making. In: Nagel, R. (Ed.), SIEDS '15: IEEE Systems and Information Engineering Design Symposium, Charlottesville, Virginia, 23 – 24 April, 2015. IEEE, pp. 212–217.

Dowding, D., Randell, R., Gardner, P., Fitzpatrick, G., Dykes, P., Favela, J., Hamer, S., Whitewood-Moores, Z., Hardiker, N., Borycki, E., Currie, L., 2014. Dashboards for improving patient care: review of the literature. International Journal of Medical Informatics 84 (2), 87–100.

Drews, F.A., Diog, A., May 2014. Evaluation of a configural vital signs display for intensive care unit nurses. Human Factors 56 (3), 569–580.

Egan, M., October 2006. Clinical dashboards: impact on workflow, care quality, and patient safety. Critical Care Nursing Quarterly 29 (4), 354–361.

Endsley, M.R., March 1995. Toward a theory of situation awareness in dynamic systems. Human Factors: The Journal of the Human Factors and Ergonomics Society 37 (1), 32–64.

Fitzgerald, M., Farrow, N., Scicluna, P., Murray, A., Xiao, Y., Mackenzie, C.F., 2008. Challenges to real-time decision support in health care. In: Henriksen, K., Battles, J.B., Keyes, M.A., et al. (Eds.), Advances in Patient Safety: New Directions and Alternative Approaches. Culture and Redesign, vol. 2. Agency for Healthcare Research and Quality, Rockville, MD.

Heath, C., Luff, P., March 1992. Collaboration and control: crisis management and multimedia technology in London underground line control rooms. Computer Supported Cooperative Work 11 (1–2), 69–95.

Hughes, J., King, V., Rodden, T., Anderson, H., 1994. Moving out of the control room: ethnography in system design. In: Smith, J.B., Smith, F.D., Malone, T.W. (Eds.), CSCW '94: Proceedings of the 1994 ACM Conference on Computer-Supported Cooperative Work, Chapel Hill, North Carolina, 22 – 26 October, 1994. ACM Press, New York, pp. 429–439.

Institute of Medicine, 2012. Health It and Patient Safety: Building Safer Systems for Better Care. The National Academies Press, Washington, DC.

Koch, S.H., Weirb, C., Westenskowb, D., Gondan, M., Agutter, J., Haar, M., Liu, D., Görges, M., Staggers, N., August 2013. Evaluation of the effect of information integration in displays for ICU nurses on situation awareness and task completion time: a prospective randomized controlled study. International Journal of Medical Informatics 82 (8), 665–675.

Kramer, H.S., Drews, F.A., 2016. Checking the lists: a systematic review of electronic checklist use in health care. Journal of Biomedical Informatics. http://dx.doi.org/10.1016/j.jbi.2016.09.006.

Kusunoki, D.S., Sarcevic, A., Weibel, N., Marsic, I., Zhang, Z., Tuveson, G., Burd, R.S., 2014. Balancing design tensions: iterative display design to support ad hoc and interdisciplinary medical teamwork. In: Schmidt, A., Grossman, T. (Eds.), CHI '14: Proceedings of the 2014 ACM Conference on Human Factors in Computing Systems, Toronto, Canada, April 26 – May 1, 2014. ACM Press, New York, pp. 3777–3786.

Kusunoki, D.S., Sarcevic, A., Zhang, Z., Burd, R.S., 2013. Understanding visual attention of teams in dynamic medical settings through vital signs monitor use. In: Lampe, C., Terveen, L. (Eds.), CSCW '13: Proceedings of the 2013 ACM Conference on Computer Supported Cooperative Work and Social Computing, San Antonio, Texas, February 23–27, 2013. ACM Press, New York, pp. 527–540.

Kusunoki, D.S., Sarcevic, A., Zhang, Z., Yala, M., February 2015. Sketching awareness: a participatory study to elicit designs for supporting emergency medical work. Computer Supported Cooperative Work 24 (1), 1–38.

Lee, S., Tang, C., Young Park, S., Chen, Y., 2012. Loosely formed patient care teams: communication challenges and technology design. In: Grudin, J., Mark, G., Riedl, J. (Eds.), CSCW '12: Proceedings of the 2012 ACM Conference on Computer-Supported Cooperative Work, Seattle, Washington, 11–15 February, 2012. ACM Press, New York, pp. 867–876.

Levine, W.C., Meyer, M., Brzezinski, P., Robbins, J., Lai, F., Spitz, G., Sandberg, W.S., 2005. Usability factors in the organization and display of disparate information sources in the operative environment. In: Proceedings of AMIA Annual Symposium, Washington, DC, 22–26 October, 2005. American Medical Informatics Association, Bethesda, MD, p. 1025.

Luten, R., Wears, R.L., Broselow, J., Croskerry, P., Joseph, M.M., Frush, K., August 2002. Managing the unique size-related issues of pediatric resuscitation: reducing cognitive load with resuscitation aids. Academic Emergency Medicine 9 (8), 840–847.

Meyer, M., Levine, W.C., Brzezinski, P., Robbins, J., Lai, F., Spitz, G., Sandberg, W.S., 2005. Integration of hospital information systems, operative and peri-operative information systems, and operative equipment into a single information display. In: Proceedings of AMIA Annual Symposium, Washington, DC, 22–26 October, 2005. American Medical Informatics Association, Bethesda, MD, p. 1054.

Muller, M.J., 1993. PICTIVE: Democratizing the dynamics of the design session. In: Schuler, D., Namioka, A. (Eds.), Participatory Design: Principles and Practices. Erlbaum Associates Inc., Hillsdale, NJ, pp. 211–239.

O'Hara, K., Gonzalez, G., Penney, G., Sellen, A., Corish, R., Mentis, H., Varnavas, A., Criminisi, A., Rouncefield, M., Dastur, N., Carrell, T., June 2014. Interactional order and constructed ways of seeing with touchless imaging systems in surgery. Computer Supported Cooperative Work 23 (3), 299–337.

Pratt, W., Reddy, M.C., McDonald, D.W., Tarczy-Hornoch, P., Gennari, J.H., April 2004. Incorporating ideas from computer-supported cooperative work. Journal of Biomedical Informatics 37 (2), 128–137.

Reddy, M.C., Pratt, W., Dourish, P., Shabot, M.M., December 2002. Sociotechnical requirements analysis for clinical systems. Methods of Information in Medicine 42 (4), 437–444.

Rogers, Y., July and August 2011. Interaction design gone wild: striving for wild theory. Interactions 18 (4), 58–62.

Salazar, A., Tyroch, A.H., Smead, D.G., October 2011. Electronic trauma patient outcomes assessment tool: performance improvement in the trauma intensive care unit. Journal of Trauma Nursing 18 (4), 197–201.

Sarcevic, A., Ferraro, N., March 2017. On the use of electronic documentation systems in fast-paced, time-critical medical settings. Interacting with Computers 29 (2), 203–219.

Sarcevic, A., Marsic, I., Lesk, M., Burd, R.S., 2008. Transactive memory in trauma resuscitation. In: Begole, B., McDonald, D.W. (Eds.), CSCW '08: Proceedings of the 2008 ACM Conference on Computer-Supported Cooperative Work, San Diego, CA, 8–12 November, 2008. ACM Press, New York, pp. 215–224.

Sarcevic, A., Weibel, N., Hollan, J.D., Burd, R.S., 2012. A paper-digital interface for information capture and display in time-critical medical work. In: Proceedings of the 6th Int'l Conference on Pervasive Computing Technologies for Healthcare – Pervasive Health, San Diego, California, May 21–24, 2012, pp. 17–24.

Sarcevic, A., Zhang, Z., Marsic, I., Burd, R.S., 2016. Checklist as a memory externalization tool during a critical care process. In: Proceedings of AMIA Annual Symposium, Chicago, IL, 12–16 November, 2016. American Medical Informatics Association, Bethesda, MD, pp. 1080–1089.

Sebastian, K., Sari, V., Yu Loy, L., Zhang, F., Zhang, Z., Feng, M., 2012. Multi-signal visualization of physiology (MVP): a novel visualization dashboard for physiological monitoring of traumatic brain injury patients. In: Engineering in Medicine and Biology Society (EMBC), 2012 Annual International Conference of the IEEE. IEEE, pp. 2000–2003.

Sommerville, I., Rodden, T., Sawyer, P., Bentley, R., Twidale, M., January 6, 1993. Integrating ethnography into the requirements engineering process. In: Proceedings of IEEE International Symposium on Requirements Engineering. IEEE, Sand Diego, CA, pp. 165–173.

Starmer, J., Giuse, D., 2008. A real-time ventilator management dashboard: toward hardwiring compliance with evidence-based guidelines. In: Proceedings of AMIA Annual Symposium, Washington, DC, 8–12 November, 2016. American Medical Informatics Association, Bethesda, MD, pp. 702–706.

Stone-Griffith, S., Englebright, J.D., Cheung, D., Korwek, K.M., Perlin, J.B., May 2012. Data driven process and operational improvement in the emergency department: the ED dashboard and reporting application. Journal of Healthcare Management 57 (3), 167–180.

Tohidi, M., Buxton, W., Baecker, R., Sellen, A., 2006. Getting the right design and the design right: testing many is better than one. In: Grinter, R., Rodden, T., Aoki, P., Cutrell, E., Jeffries, R., Olson, G. (Eds.), CHI '06: Proceedings of the SIGCHI Conference on Human Factors in Computing Systems, Montréal, Québec, Canada, 22–27 April, 2006. ACM Press, New York, pp. 1243–1252.

Vinen, J., November 2000. Incident monitoring in emergency departments an Australian model. Academic Emergency Medicine 7 (11), 1290–1297.

Waitman, L.R., Phillips, I.E., McCoy, A.B., Danciu, I., Halpenny, R.M., Nelsen, C.L., Johnson, D.C., Starmer, J.M., Peterson, J.F., July 2011. Adopting real-time surveillance dashboards as a component of an enterprisewide medication safety strategy. The Joint Commission Journal on Quality and Patient Safety 37 (7), 326.

The Recording and Reuse of Psychosocial Information in Care

Xiaomu Zhou[1], Mark S. Ackerman[2], Kai Zheng[3]

[1]Northeastern University, Boston, MA, United States; [2]University of Michigan, Ann Arbor, MI, United States; [3]University of California – Irvine, Irvine, CA, United States

1. INTRODUCTION

The strategic role that health information technology (HIT) plays in enabling the health care reform in the United States, combined with the ongoing national debate on how HIT should be used "meaningfully" to achieve the desirable transformative change, has created a critical need for research studies that contribute to a better understanding of how to utilize electronically available data for constructive, cooperative use and reuse. While electronic health record (EHR) systems provide tremendous promise for improving quality of care and controlling soaring costs, a large body of literature has noted the cumbersome usability of these systems, including numerous unintended adverse work-related and care-related consequences (e.g., Heath and Luff, 1996).

Furthermore, increasingly, doctors have to cope with patients' chronic illnesses, which affect a patient personally and socially over time beyond the disease-specific medical symptoms and treatments (Kutner et al., 1999). For example, there have been an increasing number of patients who demonstrate various kinds of pain issues, many of which are caused by, or contribute to, serious psychosocial problems they bear in life. This trend requires doctors to acquire a complete view of a patient's history in order to make informed treatment decisions.

Unfortunately, it has been shown that a patient's history can be poorly documented in an EHR system (Heath and Luff, 1996). Through this field-based study, we aimed to explore how information is used and documented to support medical work, how it is reused across a patient's multiple care episodes, and how an improved understanding of doctors' information practices could inform more accommodating and usable EHR designs. The findings explicate the dichotomized purpose of medical records, as both a representation of medical work to facilitate real-time activities (i.e., practice centered) and a representation of the

133

patient to support long-term information reuse (i.e., patient centered). In addition, this study contributes to health informatics research and practice by highlighting several key function-alities that have been missing from current designs.

This chapter is based on a 2-year field study at a large teaching hospital where the first author shadowed the routine patient care practice of over 24 physicians and residents. Data were collected in 2008–09, and reports on the use of e-Care, the system used at the time. Unless noted here, the findings still carry into current practice.

In this chapter, we examine how doctors acquire, document, and use information across multiple episodes of patient care with special attention paid to how they cope with a patient's psychosocial experience. In this study, we define psychosocial information as a patient's psychological and social issues in her illness experience. With this focus, we explored (1) under what circumstances doctors choose to document psychosocial informa-tion and what kinds of psychosocial information they choose to document and (2) how this information, or more likely its absence, affects a patient's treatment plan and subsequently the effectiveness of care. Too often the psychosocial information required to understand the patient's situation or motivations is not sufficiently documented in the EHR to be of subsequent use. This is not trivial. For instance, according to the US Substance Abuse and Mental Health Service Administration, nine percent of the US population aged 12 years or older, or 22.3 million people, were classified with substance dependence or abuse issues in 2007. Such issues could be more effectively treated by making full use of psychosocial information.

In the remaining sections of this chapter, we first review the relevant literature that serves as the guiding framework for our research. Next, we describe our field site and data collec-tion, followed by several representative patient cases describing doctors' information prac-tice. We conclude with a discussion of insights that this research helps generate into medical professionals' information behavior as well as the implications for improving the design of current HIT systems to support a better representation of medical work.

2. LITERATURE REVIEW

A hallmark of human-computer interaction and computer-supported cooperative work (HCI/CSCW) and health informatics research has been the analysis of the gap between representations of work and the work they represent (e.g., Reddy and Dourish, 2002). In health care, for example, Bossen (2006) studied a prototype EHR system constructed according to a Danish national EHR standard. The system was found to only partially support clinical work, which was largely attributable to the model used in the standard deviating from how clinical work is actually performed. Similarly, Niazkhani et al. (2009) reported that the overly simplistic representation models underlying current medication ordering systems led to severe interference with, rather than facilitation of, the actual medical work. Furthermore, Fitzpatrick (2004) showed that in reality, clinicians often tailor, re-present, and augment clinical information according to their roles and prefer-ences, which is not adequately supported in the current EHR design. Finally, researchers have demonstrated that the flexibility that allows patient records to be provisional, infor-mal, or private could facilitate care delivery (Hardstone et al., 2004) and patient hand-off processes (Engesmo and Tjora, 2006; Zhou et al., 2009). Such "informality" of documentation is generally not available in the HIT systems seen to date.

A separate but related stream of HCI/CSCW research attempts to understand the function of medical records in supporting medical work. Berg (1997) referred to medical records as a formal tool or system that embed representations describing medical workplace and activities. He argued that through clinicians' reading and writing in their patient care activities, medical records play a fundamental and constitutive role in supporting medical practice (Berg, 1996). In studying e-prescribing applications, Gorman et al. (2003) argued that HIT systems are useful only when their designs accommodate and facilitate clinical activities as a multidisciplinary collaboration effort and fit better into the larger system of patient care.

Part of this stream concerns the question whether medical records should be conceptualized as process centered (i.e., organized around a medical facility's work processes) or as patient centered (i.e., organized around the patient's disease descriptors and health conditions). For instance, Østerlund (2004) depicted medical records acting like a "map and itinerary to guide clinicians' work," and thus he favored the process centered organization. As we will see, this distinction is critical to the design of medical record systems.

Finally, to examine the appropriateness (accuracy and comprehensiveness) of representations of medical work in the context of medical records design, we found the concept of *trajectory*, a term that Strauss and colleagues (Glaser and Strauss, 1967; Strauss, 1993; Strauss et al., 1997) first coined, useful in our analysis. According to Strauss, a "clinical course" differs from an "illness trajectory." The clinical course describes what has happened since the patient's admission, such as reasons for the admission, medically meaningful symptoms, and diagnostic results and treatment plans; whereas an illness trajectory refers "not only to the physiological unfolding of a patient's disease but also to the total organization of work done over the course, plus the impact on those involved with that work and its organization" (Strauss et al., 1997, p. 8). The difference between a specific clinical course and an illness trajectory, as we show in the later sections of this chapter, is useful in understanding doctors' information practices and the role of medical records in supporting (or hindering) such practices.

3. ABOUT THE STUDY

We collected the field data by observing a general internal medicine team. This team was selected because its work is in line with our primary research interest, long-term use of medical information. The team provides service to patients who often have chronic episodes of their illness across their adult lifespan and come to the hospital when they experience a flareup or other acute situations. Observing this service's work would thus provide rich data on information reuse issues from a long-term perspective.

3.1 Participants

The team, called the Medicine Howard (MH) service, is one of four general medicine services in the department of internal medicine. It usually consists of one attending physician (referred to as an *attending* in this chapter), one or two second-year residents (*residents*), and two first-year residents (*interns*). Occasionally the team hosts one medical school student. Each month, one of the four senior physicians who belong to the MH service supervises the residents and interns, who also rotate through the service. During our 9 months of observation, three attendings, nine residents, twelve interns, and two medical students participated in our study. In addition, we observed the work of another team periodically in order to gain a broader understanding of doctors' work.

3.2 Data and Data Collection

This study consists of largely field-based observations augmented by the examination of patients' medical records in the EHR system, in this context, e-Care. The first author performed the field observations. She shadowed doctors' overall work, typically from 3 to 5 h each time. On two occasions, she shadowed the team throughout their on-call day, that is, 30 consecutive hours in the hospital. The observational activities involved following the teams' patient care activities, asking clarification questions, tracking critical incidents, and taking field notes. Between observations, the researcher reviewed patient records and working documents. In addition, whenever appropriate, the researcher also asked to look at personal rounding sheets in order to understand how the attendings, residents, and interns assembled information. She also attended the educational meetings and lectures organized by the attendings. She was even invited to the team social events, such as the dinner party when a rotation ended.

The first author was also granted access to the e-Care system, so she could conduct an in-depth analysis of relevant research issues captured in the medical records. The e-Care system, used at the time of the study, was a web-based medical records application providing authorized users real-time access to patient data. It integrated, to a limited extent, information residing in other electronic systems of the hospital, such as Emergency Department (ED) diary notes, medication orders, laboratory work, and data from radiology, cardiology, neurology, registration, and other special care units. It included clinical notes from doctors, nurses, and other clinical personnel (e.g., admission notes, progress notes, nursing notes, discharge summaries, and social worker notes).

Our investigation began with an examination of the overall work of the MH team, which spans a wide range of activities including patient admission, initial diagnostic interviews, morning rounds, post-rounds group discussions, generating notes, providing medications, team meetings, sign-out process, and so on. Our attention was soon attracted to the information assembling process, particularly when the team admitted new patients, and to the morning rounds immediately after an on-call day, when diagnoses and treatments were intensively discussed among the team members. The first author observed a total of 260 patient room visits during morning rounds, among which 104 were the first visit after the patients were admitted. Additionally, over 70 patients' records (30 with substantial psychosocial issues) were reviewed with a special focus on the doctors' comprehensive assessments of each of the patient cases.

For the study reported in this paper, we extracted the portions from our field observational notes related to information seeking and assembling activities that occurred immediately following patient admission. We identified information use issues from a social/symbolic interactionism perspective (Glaser and Strauss, 1967; Strauss, 1993) and paid close attention to the occurrence of psychosocial issues in the work of care. We then investigated whether the psychosocial information was, or was not, documented in e-Care by reviewing the corresponding patient records. Field notes and medical records were used to corroborate one another during the data analysis process.

Any cases described in this chapter are summarized from the field notes and examination of patient records retrieved from the EHR. All data, including names and the site's name, have been anonymized.

4. DOCTORS' WORK

Over 80% of the patients on the MH service are transferred from the ED at the hospital. The remaining patients are referred from ambulatory care. Patients usually stay on this service for 3–4 days on average, with a wide range from a 1-day stay to over a month-long hospitalization. MH takes patients whose symptoms do not fit into any of the clearly defined special service teams (e.g., cardiovascular, gastroenterology, hematology, and oncology); thus, the MH patient pool covers a range of profiles including arthritis, asthma, diabetes, hypertension, and heart disease. Many patients who have chronic nonmalignant pain issues are also often assigned to this service.

This situation requires the MH team to deal with a mixture of complicated issues. The residents of this team usually arrive at the hospital early enough to conduct individual visits with their patients and prepare for the morning rounds. Morning rounds start between 7 and 8 a.m., and they usually last 2 to 3 h depending on how many new patients have been admitted. After morning rounds, the residents always talk with each intern again in order to make sure that the treatment and entire care plan will be carried out and done on schedule. Doctors then spend the rest of the day working on their own, although interacting (via phone) with specialty teams, family members, primary care doctors, social workers, discharge planners, and nurses also constitutes a large part of their work.

In the remaining part of this section, we use illness trajectory as a guiding analytical concept to describe and interpret our findings along two major lines: *information use* and *documentation*. First, we present briefly how medical information is acquired, assembled, and used in a general illness trajectory (case 1). Then, we describe how doctors process psychosocial information with three illustrative cases: (1) where a psychosocial issue occurred in a trajectory but was not documented by doctors (case 2); (2) where a psychosocial issue, supported by definitive evidence, was communicated among doctors (and with other medical professionals) and was subsequently documented in e-Care (case 3); and (3) where in certain circumstances psychosocial information was judiciously documented and used (case 4). While presenting these cases, we highlight how the absence of psychosocial information (i.e., the missing representation) may have had an impact on quality of patient care and costs.

4.1 Information Acquiring and Assembling

Information seeking and assembling takes place simultaneously in the process when MH admits new patients, conducts diagnostic interviews, and evaluates a patient during morning rounds. The most intensive information seeking and assembling occurs right after admitting a patient.

The work starts with a paging text from the ED or the admitting unit to the resident, which includes a possible diagnosis. The resident immediately makes a quick assessment based on the ED diary notes in e-Care to decide whether this patient is appropriate for the MH service. Next, the resident may briefly talk with the ED doctor and then assign this patient to one of the interns. When a patient is referred to the hospital, the resident often expects a primary care physician's note in the e-Care system. Both the attending and the resident(s) supervise the interns, but ultimately it is the interns who are responsible for generating the medical records (admission notes, progress notes, treatment plan, discharge document, and so on), which will be subsequently reviewed by the residents and revised (if necessary) and signed by the attending doctor.

A doctor rarely goes to see a patient for a diagnostic interview without careful preparation. She needs to have a relatively convincing idea of what is going on (e.g., several possible causes) with this patient. In some cases, a patient comes to the hospital for a chronic illness flare-up that has been treated before in this hospital. If the laboratory results, vital signs, and other measures are very consistent with what has been observed before, the anticipated trajectory can be very routine and predictable. For other patients, however, the resident and interns may not be able to make sense of the case based on the patient's symptoms and performance and their possible causes. In such cases, the doctors use additional information sources. The following case demonstrates this.

CASE 1

A patient was transferred from another hospital as an emergency case. He has past medical history with post kidney transplant and hypertension. Recently he took a vacation to Honduras for a scuba diving trip. After he flew back, he developed nausea with vomiting. In another hospital, his situation improved, but he was found to be hypoxic (i.e., low oxygen in his blood). Based on a concern for him as a kidney transplant patient, the patient was transferred to this hospital for further evaluation.

This case highlights the intense informational activities during the preparation for a diagnostic interview.

The intern reviewed the ED diary notes, laboratory test results, and the medical records sent from the outside hospital in order to prepare for meeting with the patient. She could not understand why the patient had developed decreased oxygen saturation with all vital signs and other descriptors appearing fine. After searching an online clinical information database for "hypoxic" causes, she started to examine this patient's previous records one by one in e-Care. Eventually, the intern discovered that the patient had experienced a similar condition 2 years ago but later recovered without further medical intervention. After this effort, the intern conducted the diagnostic interview.

Diagnostic interviews often take place shortly after a patient is admitted to the MH service. The resident and the intern conduct independent interviews with the patient. During an interview, 14 categories of questions will be asked, each relating to one part of the human body system. The interview usually goes in a matter-of-fact style, Q&A fashion, and at fast speed. However, because the doctors want to investigate information about not only symptoms but also about the patient's past medical history, family and social history, and lifestyle (i.e., the entire context of the illness experience, which often includes sensitive psychosocial information), a diagnostic interview may lead to a very emotional reaction. For instance, when one female patient was asked about her pregnancy history, a previous miscarriage caused her to burst into tears.

Doctors often have to learn skills to deal with patients who present with problematic behaviors. For instance, the interns and residents often share tricks, which they name "distractible components," to discover whether a patient is truly suffering pain or just demanding a controlled substance. Patients with substance abuse histories often present at the ED complaining of severe "abdominal pain," since it is expensive to screen out all potential causes. Inconsistent reactions to each physical assessment are considered to be faking the symptoms. The team members often share information among themselves verbally about those patients who are likely to fake their symptoms. This observation is similar to that by Strauss et al. (1997) that moral judgments are very frequent and severe in emergency rooms.

Finding out about a patient is a process of information sharing, sense-making, decision-making, education, and training. For instance, patients often tell different doctors different stories or stories of more or less depth about their illness experience, particularly about the psychosocial issues in their lives. Morning rounds provide an opportunity for the team to piece together the information and gain a better understanding of their patients. In a patient's room during morning rounds, psychosocial information is often acquired through talking with family members individually and with other caregivers, such as home visiting nurses.

As searching and acquiring information develops along a trajectory, assembling the information takes place simultaneously. Each doctor has her version of the rounding sheet, whether a structured template or a piece of blank paper. Each patient gets one sheet. This rounding sheet appears to be the most important working document for doctors to carry around in their pockets. The rounding sheet will be manually filled in with a patient's demographic information, emergency contact, history of present illness, past medical/surgery history, ongoing medication, family/social history, newest radiology/laboratory results, and so on.

5. DOCUMENTING HEALTH CARE INFORMATION

A great deal of information is generated during the process of a developing trajectory. What information do doctors document? How do they write a patient's information, especially psychosocial, into the medical records?

The e-Care electronic patient records system used in the study hospital is a Web-based application that allows doctors, nurses, and other clinicians to generate free-text notes, including admission notes, progress notes, discharge documents, nursing notes, social worker notes, and special consulting notes. All documents are arranged chronologically; and at the time of the study, there was no keyword search.

An admission note includes predefined categories of information including a patient's chief complaint, detailed history of present illness, past medical and surgery history, family and social history, and the assessment and plan. Among the various notes, the admission note contains the most comprehensive information about a patient and is the first document that the service team provides. It is used throughout the trajectory not only by the team itself but also by nurses and other clinicians as both a source of baseline information and a guide for the work of care.

Among various categories of information in an admission note, several are matter-of-fact and straightforward, but others can be questionable and sometimes require careful wording (see later cases in this chapter). For instance, "family history" usually records whether family members have a similar or related disease; "social history" should include any information about the patient's living situation, occupation, or any other aspects of the patient's life that may be clinically significant to the patient's problem. "Social history" is supposed to contain information such as where and with whom the patient lives, employment, social support, activities, habits, insurance coverage, feelings of anxiety or depression, visits to psychiatry or social workers, and ability to care for oneself (if elderly). All of this information will tell a doctor how a patient manages her illness in her social situation. However, according to one attending doctor, in practice, the "social history" has deteriorated to include only habits such as smoking, drinking, and illegal drug use.

In the "history of present illness" section, doctors write in free-text how a patient presents at the hospital, various symptoms, and other phenomena they observed or stories they investigated via a diagnostic interview with the patient and discussion with her family members. At the end of an admission note, the "assessment and plan" should document a doctor's rational thinking, that is, their interpretation of the patient case and why this patient should receive this particular treatment. A good admission note should address the issues clearly and provide a convincing rationale for the treatment plan. However, the critical thinking or supporting evidence is often missing, leaving later doctors to wonder why the patient received an intervention during the previous episode. Psychosocial issues (if documented) often appear in the "history of present illness" and the "assessment and plan" sections.

As psychosocial information is often considered to be subjective and is often vaguely defined or perceived differently by different care providers, the handling of such information magnifies the gap between the work, the patient, and the representation (i.e., medical record). In the following sections, we describe three cases that illustrate how doctors cope with patients' psychosocial issues; how they interpret, use, and document psychosocial information; and, how the breakdown in the representation can potentially affect clinician performance, quality of care, and costs.

5.1 Psychosocial Information, but Only in "Talk"

Consider the following example:

CASE 2

A 36-year-old female patient with history of hypertension and anxiety disorder presented at the ED with the complaint of chest pain. She was assigned to the MH service and was waiting for a bed. Upon arriving at the ED, Kristine, the MH resident, overheard a nurse say that this patient showed up at the ED every few days. Often, the patient received an intravenous (IV) infusion (with a controlled substance) and then was discharged. On several occasions, she was hospitalized for further evaluation, so she could get more pain medications. The laboratory/radiological data did not reveal anything clinically significant. When Kristine communicated this case to her attending, the attending became outraged and immediately led the entire team to the ED. The attending speculated that the patient was manipulating her symptoms to gain access to a controlled substance. The attending confronted the ED doctor. Eventually, the patient was discharged from the ED as requested by the MH service.

This was a problematic care trajectory, which ended with the attending's interaction with the ED doctor. However, the record did not document the conflicting understandings of the attending and the ED doctor nor any of the patient's problematic behavior. It may be speculated that when this patient arrives at the hospital again, she may be admitted to a different service or even to the same service when the attending, residents, and interns are different (due to periodical rotations). For this case, although the psychosocial issue emerged as a main concern, it still did not seem legitimate enough to be documented in the record. As one resident stated, "You never know for sure."

Patients demonstrating pain symptoms are prevalent in this study site. Yet, e-Care did not provide a systematic means for the medical teams to formally capture this information as part of a patient's record or perhaps better, in informal documentation (as noted in Hardstone et al., 2004), so that this information can be noted down and shared across care episodes. This points to missing technical capability for supporting this type of long-term information reuse. Whether or not to record this sensitive information and how to record it is largely left up to each individual doctor. Many other psychosocial issues critical to understanding a patient's needs and motives are also shared only verbally. This leaves the next care team in an information vacuum and requires the repetition of time-consuming investigations in complicated patient conditions.

5.2 Psychosocial Information in the Record, but When?

Under certain circumstances, psychosocial information may be documented in the formal representation. However, its importance may not be immediately recognized by every member of the medical team. The psychosocial information is largely passed along orally in the beginning of a patient's illness. Perhaps it will be eventually captured in e-Care, but this may not occur for a long time. In the following case, it happens that a patient resorts to violent behavior, and then doctors have "hard evidence" to note in the record.

CASE 3

(All quotes are from doctors' notes in e-Care.) A 23-year-old woman with a history of sickle cell disease comes to the hospital ED every few days complaining of chest pain. During the last hospitalization, the patient had "significant issues with behavior." When she was told she could not have IV Benadryl (an abusable substance), "she became quite frustrated and ripped up all of her paperwork. ...She physically threatened numerous staff members and required security presence on more than one occasion." The MH service ordered full tests, then noted, "there was no evidence of acute chest syndrome demonstrated. ...It was not felt that the patient was exhibiting evidence of serious sequelae of sickle cell crisis."

The attending talked with the patient's primary care physician to put her on a chronic pain management program, which might eventually help the patient stop the drug abuse. They jointly made it very clear in the patient's discharge notes, she "should no longer get IV Benadryl and she was abusing this."

Although this case was of a similar nature to case 2, details were recorded in the e-Care system to inform others about this patient's conditions, which, if used properly, could prevent these issues from happening again.

As an aside, there is no guarantee that such information would be re-examined, since reuse is subject to visibility, incentives, and the power relationships among doctors. The next ED doctor missed the information written in the discharge notes in e-Care.

After only a few days, the patient showed up at the ED complaining of nausea, vomiting, and severe pain in her legs and back. She again demonstrated questionable behavior, refusing a chest X-ray when she did not receive IV narcotics. Then the ED doctor gave her one dose of IV Benadryl, which violated her ongoing pain management program that the attending and her primary care physician setup.

The ED routinely uses another electronic system, which records a patient's vital signs and other medically critical information but does not have a patient's detailed past medical history. If the ED doctors want, they can login to e-Care to find out a patient's past episodes, but this requires extra effort. As well, there are distinct differences in the priorities between ED doctors and floor doctors (those doctors such as the MH team). ED doctors' priorities are in treating the immediate symptoms and moving patients to floor units as quickly as possible. Floor doctors, on the other hand, not only deal with acute conditions but also need to plan for long-term care. It is not necessarily in an ED doctor's interest to face down drug abuse, as this could considerably slow down the interaction with a patient. Floor doctors, on the other hand, must do a great deal of unnecessary work for patients seeking drugs. Accordingly, there is a tension between floor doctors' desire to have ED doctors to carefully read patients' prior records and the ED doctors' incentives to ignore prior information. The lack of visibility does not help. We will return to this issue below.

So far, we have described cases where the psychosocial information was never recorded and where it was recorded late in a patient's history. Next, we examine a case where it was recorded appropriately.

5.3 Detailing Psychosocial Information in the Record

Some trajectories may be dominated by the psychosocial factors to such an extent that without those issues being in the patient's record, the necessary work cannot be accomplished. Below is a case that illustrates psychosocial information being systematically captured in the medical records from the very beginning of a trajectory (as compared to case 2 and case 3, where the psychosocial issue was never recorded or recorded only after severe events had occurred).

CASE 4 (ALL QUOTES ARE FROM THE RECORDS IN E-CARE)

Day 1: Mrs. Smith, an 81-year-old patient with a history of dementia, anemia, depression, and hypertension, presented at the ED with multiple falls. ED doctors noted the patient "had some ecchymosis (skin discoloration caused by blood) over the right side of her face.... The number of falls the patient has had over the last several days is concerning, especially given her living situation." The MH team resident Nancy and the intern John conducted diagnostic interviews separately and examined the patient carefully. They had serious concerns.

Day 2: Nancy and John reported to the attending that they called the home visiting nurse, who reported that the patient's son who lives nearby said, "Dad beats Mom." After the attending carefully examined the patient, he noted in the admission note, "It is unclear how one discrete fall could cause the variety of bruises on the patient, including the ... edema, arm bruises, and side bruises. This may be consistent with multiple falls over time because of dementia, however abuse should be considered in this case as well...." The attending pushed for a meeting with the family and to include a social worker.

Day 3–5: Various personnel were called to evaluate Mrs. Smith. Her primary physician was also informed. Diane, a practice management coordinator, phoned Adult Protective Service (APS) and the Visiting Nurse Organization (VNO). She noted in the records that the VNO expressed "their concerns of the safety in the home due to Mr. Smith's sexual advances toward the home visiting nurse." Soon, APS became involved in the case.

Surprisingly, Mrs. Smith, who was believed to be non-conversant, became more verbal, mumbling "they are mad at me" and "everyone is yelling and asking me what I am trying to do."

Day 6: A progress note noted "significant bruising over her body, concern for elder abuse. Adult protective services has been contacted, are currently investigating her case. Unsafe to go home."

Day 7–12: While all parties worked hard to investigate the problem, the family was trying to have the patient discharged to her home. Nurses noted in records that the patient had a "crying episode overnight for 5 hours".

Day 13: The meeting of all parties took place. The APS representative "discussed with the family legal actions against them for their noncooperation."

Day 16: Mrs. Smith was discharged to a nursing facility. Family may not take the patient from the nursing facility without discussing their plan first with the APS agency.

The hint of elder abuse, the psychosocial information, was noted in the records from the very first day. Along the development of the trajectory, details of elder abuse and complicated troublesome family dynamics were increasingly discovered and documented in the records. Compared to other trajectories, in which the explicitness and accountability of the psychosocial issues in the records were limited, psychosocial issues were at the core of this trajectory and this was reflected in the records.

This is a very special trajectory that highlights the complexity of the emotional work in some cases of medical care. Several issues are of note. First, it is stunning that how many details related to psychosocial issues that the MH team investigated and documented in the record. Furthermore, the attending pushed very hard on this case to get all parties involved; otherwise, Mrs. Smith might have been just treated as a normal "dementia patient fall" case.

Second, as described in the story, there are many clinical personnel (e.g., ED doctors, MH team, nurses, social workers, practice management coordinators, and the primary care physician) and several social services (e.g., APS, county sheriff, and nursing home) involved in this trajectory. Each of them had their specific role in solving medical issues (perhaps simple in this case) and social issues (extremely complicated). The hospital clinicians described their work and their understanding of the case in e-Care in real time. Information sharing was very intensive, as a coordination to collectively investigate the issue and solve the problem. In this case, the medical work of care was marginal (i.e., treating bruises), but the information work was at the very core of the entire trajectory.

Third, the patient and family members, who were fighting among themselves, were noncooperative with doctors and social services, and they complicated the trajectory by not being able to provide or by attempting to hide information. However, the information was pieced together collectively, and the doctors tried to write the consequences of each step in the records. In this case, e-Care was able to satisfy the needs of the clinical workflow and work representations in this case.

This case showed how the medical team, when they felt it appropriate, would document the psychosocial information for a patient. Clearly, this case was unusual. It highlights, nonetheless, the emphasis on the doctors' sense of "appropriateness" in determining when to document. We turn to a discussion of this next, as well as design implications from our study.

6. DISCUSSION

Our field observations reveal the need for additional consideration of psychosocial issues in medical practice. In our site, this was due to complicated patient profiles, chronic illnesses throughout patients' lifetimes, and poorly controlled pain issues.

There are three important findings from this study for medical information systems design. The first is that doctors will detail psychosocial information; however, they do not *always* document this information, as demonstrated through the differentiated handling of such information in case 2 and case 4. Why might this be the case?

We believe this is a result of the way that doctors are trained to use their documentation. Doctors are trained to look for *symptoms* first, then they think about the *causes* (based on their medical knowledge and their experiences). This is the sense-making stage and also the medical reasoning process that leads to diagnostic judgments. Finally, they need to come up with a *treatment plan*. Therefore, symptoms, possible causes, and treatment and care plans are perhaps the most important three categories of information in medical records to represent their work. These categories of information also constitute valuable information for future reuse when a patient is readmitted to the hospital.

If suspected "causes" match "symptoms" well, a trajectory will be straightforward, even though achieving it may not always be uneventful (as in case 1). In an internal medicine unit, most patients are admitted because of acute events due to chronic illness, so the "cause" is easily assumed to be medical. In case 4, the symptoms were bruises, and the cause was a "fall" (according to the family members' report). However, a single fall was not likely to cause so many bruises on her body (as the attending noted in the records), and if the bruises were caused by multiple falls, how did these falls happen? Doctors needed to provide a convincing diagnosis, so they went further. In this case, the "cause" was psychosocial, but the symptoms were medical. This was reflected in the records, where a great deal of psychosocial information was documented. In addition, the treatment could not address just medical issues. The doctors needed to prevent the abuse from happening again, so they pulled together all sources to find a suitable treatment plan.

In case 3 however, the pain drug-seeking patient had a medical issue, that is, sickle cell disease. Although she was admitted to the hospital frequently, the doctors still first looked for symptoms. The symptoms were documented in the records as "questionable behaviors" because they did not match sickle cell disease (i.e., the cause). The doctors speculated that the patient was faking the symptoms. In this scenario, the "symptoms" became psychosocial, or at least a mix of medical and psychosocial. In reviewing previous records of this patient, the doctors did not put appropriate information in her records until the most recent episode in which the patient became violent and threatened others. This became the triggering incident that provided evidence for the doctors' speculation. Lacking definitive evidence, doctors may hesitate to document such suspicions of "faked symptom" in the medical record. This is reflected in case 2, where the doctors speculated that the patient was seeking drugs but did not explicate it in the records. This missing representation of psychosocial information may eventually create severe problems, such as the incident described in case 3, where the psychosocial issue was finally brought to the medical team's attention and documented in writing. However, it may have been too late for the patient.

Indeed, this story is not extraordinary: Over the past several decades, there has been a tendency to view *all* patient-presented complaints and symptoms as curable diseases that can and should be treated within the purview of medical professionals (Gallagher and Ferrante, 2005). This view, however, often leads to an overly narrow, "medicalized" lens of health and illness that largely ignores psychosocial causes and other contributing social and economic factors. Smoking and obesity, for example, can be viewed merely as diagnosable and curable diseases and treated with nicotine substitutes and obesity drugs; however, this defocuses their behavioral and socioeconomic root causes (Lantz et al., 2006).

Medicalization is defined as "the expansion of medicine as an institution and the use of a medical lens to view human processes and behavior" (Zola, 1972). We believe it is largely this medicalized view, not the sensitivity of information, that sets the boundary of what information to be documented and what not to be. It is also this medicalized view that determines the reuse value of information in subsequent care episodes. Medicalization implies clear diagnostic tests and evidence. Oftentimes certain psychosocial information gets lost, as in case 2, because such information is not yet formally defined in medicalized terms and encompassed in the medicalization view. Such information is relegated to the "subjective", becoming less than a "medical fact".

Case 4 illustrates a rather unique case where the medical team transcended the boundary set by the medicalized view to actively seek help from other parties including social services. In this case, the symptom, "bruise," was clearly disconnected from the suspected medically relevant cause, "fall," which obliged the medical team to think out of the box to find nonmedical evidence and seek nonmedical interventions. This endeavor, however, does not always take place because such a disconnection is not always readily discernable.

Second, our findings point once again toward the need for considering the broader context of medical information systems. The information models underlying current EHR systems are mainly organized around storing and managing symptoms and treatments. For example, the Certification Commission for Health Information Technology, the accreditation body for commercial EHR products, places an exclusive emphasis in their certification criteria on whether an EHR system has the capability of capturing and managing discrete, process-oriented, and medicalized data, rather than on the meaningfulness (and cognitively coherent representations) of the data to clinicians in their patient care activities. Our study shows the need to gain legitimization for psychosocial issues in system construction and include appropriate representations in the record formats.

Third, and most importantly, as we have shown, there exists a gap between the work, the patient, and the representation, which may account for the suboptimal outcomes or adverse consequences observed such as repetitive investigations, delayed diagnoses, inappropriate treatments, unnecessary hospitalizations, and increased cost burdens for the hospital.

This broader implication raises the need to reconceptualize medical records adaptively as both a representation of medical processes and of the patient. An extensive body of literature in HCI/CSCW has been devoted to studying the issues related to the representations of work, recognizing the inherent gap between representation and the real world, and how systems should be designed to support ongoing work activities (e.g., Bossen, 2006; Schmidt, 1997). Our study points to a new perspective that representation of information may need to be constructed in adaptive forms when a singular form cannot adequately support a multiplicity of purposes, changing demands across time, and distinct priorities of the information consumers. In the medical context, while the information representation that supports medical processes—routines and procedures in day-to-day care—remains critical, what needs to be shared across multiple patient care episodes is not only the process-oriented information but also information centered around the patient's life long illness trajectory (Strauss et al., 1997)—her medical conditions and other associated psychological and social experiences. As shown in this paper, the conceptual models underlying current medical records are largely process centered, which do not accommodate this multifaceted need and hence may adversely affect medical practice and diminish the reuse value of documented patient care information. Our study represents an attempt to examine whether focusing on one model may lead to the missing of critical functionalities for the continuity of care when a patient comes back.

7. DESIGN IMPLICATIONS

Our findings provide several insights into redesigning medical information systems from a socio-technical perspective. First, it shows it is necessary to consider the organizational incentives for all of the stakeholders. In this setting, it was clear that pain medicine abuse is acerbated by the ED doctors' tendency to move patients through as quickly as possible. Technically, providing all doctors convenient access to critical information about patients is also important. For example, many US states have now implemented patient registry systems that maintain a comprehensive list of patients' prescriptions. This list is accessible to licensed physicians free of charge; integrating such information directly into EHRs could help address the issue of information visibility.

Second, this study highlights the need for a technical capability of documenting psycho-social information—this would allow clinicians to consider the "whole" of a patient. This psychosocial information is often perceived as "informal" when definitive evidence is not yet available. EHR systems, such as e-Care, are not only designed to support care processes but also to focus on the capture of billable, "medicalized" information. As we have seen, EHR systems lack the ability to document and use "informal" and provisional information, as argued in Hardstone et al. (2004), particularly the information that sheds important light on patients' psychosocial issues. In our site, such information was then communicated only verbally and therefore not communicated to the next team effectively.

Third, our study also suggests the importance of considering information's long-term use more broadly. At this site, understanding the patient from a long-term perspective is far too difficult due in part to the technical difficulties of reusing patients' medical records across multiple episodes. When information reuse occurs within an episode, clinicians need explanatory details to help them understand the current trajectory; when it occurs across episodes, they need to know key issues about the patient. This was reflected in case 1 when the doctor had to read an immense volume of past records, line by line, in order to identify the information she needed. This reiterates the need for mindful consideration when constructing medical records for multiple purposes. An EHR system should be designed to facilitate the clinical work in a nuanced way (i.e., process-centered representation) while *simultaneously* preparing information of high value about the patient for long-term reuse (i.e., patient-centered representation).

8. CONCLUSION

This field-based study describes doctors' use and documentation of medical information, in particular, psychosocial information. We found that doctors documented a considerable amount of psychosocial information in the EHR. Yet, we also noted that such information was only recorded selectively, with a "medicalized" view of appropriate information being a key contributing factor. As well, our study showed how problematic and missing representations of a patient seriously affect work activities for the medical team and for a patient's care, especially for chronic conditions. We accordingly suggest that electronic systems in health care should be designed to support both representations of medical processes and of the patient.

Acknowledgments

This work was supported in part by a University of Michigan Rackham Barbour Scholarship, the National Science Foundation (0325347), and the National Institutes of Health (UL1RR024986). The authors thank the internal medicine teams at the study site for their incredible support and patience.

References

Berg, M., 1997. On distribution, drift and the electronic medical record. In: Proceedings of the European Conference on Computer-Supported Cooperative Work (CSCW1997). ACM, New York, pp. 141–156.

Berg, M., 1996. Practices of reading and writing: the constitutive role of the patient record in medical work. Sociology of Health & Illness 18 (4), 499–524.

Bossen, C., 2006. Representations at work: a national standard for electronic health records. In: The Proceedings of the ACM Conference on Computer-Supported Cooperative Work (CSCW2006). ACM, New York, pp. 69–78.

Engesmo, J., Tjora, A., 2006. Documenting for whom? New Technology, Work and Employment 21 (2), 176–189.

Fitzpatrick, G., 2004. Integrated care and the working record. Health Informatics Journal 10 (4), 291–302.

Gallagher, E., Ferrante, J., 2005. Medicalization and social justice. Social Justice Research 1 (3), 377–392.

Glaser, B., Strauss, A., 1967. Discovery of Grounded Theory: Strategies for Qualitative Research. Aldine Publishing, New York.

Gorman, P., et al., 2003. Order creation and communication in healthcare. Methods Informatics Medicine 42 (4), 376–384.

Hardstone, G., et al., 2004. Supporting informality. In: In the Proceedings of the ACM Conference on Computer-Supported Cooperative Work (CSCW2004). ACM, New York, pp. 142–151.

Heath, C., Luff, P., 1996. Documents and professional practice: 'Bad' organisational reasons for 'good' clinical records. In: The Proceedings of the ACM Conference on Computer-Supported Cooperative Work (CSCW 1996). ACM, New York, pp. 354–363.

Kutner, J.S., et al., 1999. Information needs in terminal illness. Social Science & Medicine 48 (10), 1341–1352.

Lantz, P., et al., 2006. Health policy approaches to population health. Health Affairs 26 (5), 1253–1257.

Niazkhani, Z., et al., 2009. The impact of computerized provider order entry (CPOE) systems on inpatient clinical workflow: a literature review. Journal American Medical Informatics Association 16 (4), 539–549.

Østerlund, C., 2004. Mapping medical work: information practices across multiple medical settings. Journal for Information Studies 3, 35–44.

Reddy, M., Dourish, P., 2002. A finger on the pulse. In: The Proceedings of the ACM Conference on Computer-Supported Cooperative Work (CSCW 2002). ACM, New York, pp. 344–353.

Schmidt, K., 1997. Of maps and scripts. In: The Proceedings of the ACM Conference on Supporting Group Work (GROUP1997). ACM, New York, pp. 138–147.

Strauss, A., 1993. Continual Permutations of Action. Aldine de Gruyter, New York.

Strauss, A., et al., 1997. Social Organization of Medical Work. Transaction Publishers, New York.

Zhou, X., Ackerman, M., Zheng, K., 2009. I just don't know why it's gone: maintaining informal information use in inpatient care. In: The Proceedings of the ACM Conference on Human Factors in Computing Systems (CHI2009). ACM, New York, pp. 2061–2070.

Zola, I.K., 1972. Medicine as an institution of social control. Sociological Review 20 (3), 487–504.

Challenges for Socio-technical Design in Health Care: Lessons Learned From Designing Reflection Support

Michael Prilla[1], *Thomas Herrmann*[2]
[1]Clausthal University of Technology, Clausthal-Zellerfeld, Germany; [2]Ruhr University of Bochum, Bochum, Germany

1. INTRODUCTION: DESIGNING SUPPORT FOR COLLABORATIVE REFLECTION IN HEALTH CARE

This chapter is based on the description and analysis of a case study in designing collaborative reflection support conducted together with a German neurological hospital. At the hospital we worked with a ward dealing with stroke patients. The ward was run by two senior physicians, who coordinated six to eight assistant physicians and the nurses of the ward. All members of staff on the ward were integrated tightly into research and development on supporting their reflection from the beginning of the study. As an early result of this integration, the topic of supporting physicians in learning about their conversations with relatives of patients was chosen as a theme for our work. Physicians told us that they perceived a need to improve their skills in conducting these conversations and that the current lack of skills created emotional stress and a bad reputation for the ward and the hospital. Similar case studies on this topic have been conducted in nursing homes and in public administration settings (Prilla, 2015; Prilla et al., 2015), and these case studies back up our findings described here. This work was done as part of a large European Commission project.fn11[1]

In our work, we faced several challenges related to the health care domain that made the design of socio-technical support for reflection at work harder. These challenges included aspects of technology adoption, alignment to structures and processes in health care, technical

---

[1] This work was done as part of the MIRROR project funded by the European Commission in FP 7. The MIRROR projects aim at supporting reflection in various settings, stages, and levels. More information can be found at http://www.mirror-project.eu/.

support for solutions, and many more, which will be reported in this chapter. The questions related to these challenges that we want to contribute to are as follows: (1) *What are the specific socio-technical design challenges related to health care?* and (2) *How can we deal with them?* If some of the challenges we faced are typical for health care settings, they provide a good basis to reflect on practices for dealing with them. Below we first give an overview of the context of our work and how we tackled the design process. Then we investigate some of the challenges we faced and give initial answers to these questions by comparing the work in our cases to previous work on similar challenges. Because we ran a similar study with similar results in a nursing home, the outcomes of this comparison are applicable to a wide range of health care work places.

2. BACKGROUND: SUPPORTING REFLECTIVE LEARNING IN HEALTH CARE

Reflection is a means for informal learning at work, which consists of looking back at past experiences, reassessing them in the light of current experiences or knowledge, and drawing conclusions from this process for future behavior (Boud, 1985). It has been called a necessary attitude for people in modern workplaces (Schön, 1983) and a mind-set to be cultivated and spread (Reynolds, 1999). Information and communication technology (ICT) can play multiple roles in supporting reflection, ranging from the provision of data to reflect upon to sustaining a systematic reflection process and, ultimately, to spreading results.

Learning has always been an important topic for health care professionals, who need to constantly be up to date with current medical practice, legislation, and needs like documentation for accounting. It does not come as a surprise, therefore that reflection has been found to be a common and well-established practice in health care workplaces (Forneris and Peden-McAlpine, 2006; Mann et al., 2009; Teekman, 2000), particularly concerning diagnosis and medical practice work (Mann et al., 2009; Teekman, 2000) as well as training (Forneris and Peden-McAlpine, 2006; Mann et al., 2009). Designing ICT support for such reflection needs to be understood as socio-technical design, as it involves the need to carefully embed supportive technology into health care work and influences the daily practices of health care workers, as our work showed (Prilla, 2015).

The work presented here focuses on another area of learning for health care professionals, specifically their interaction with patients and relatives. Relatives of patients are often hard to talk to. Therefore, it can be a stressful and burdening activity, as it often involves conveying bad news (Maynard, 2003) or dealing with emotionally stressed conversation partners (Prilla et al., 2013). Such conversations are often conducted in an ad hoc manner, without the possibility for professionals to prepare (Delvaux et al., 2005; Pennbrant, 2013), and experience is required to handle such difficult situations successfully (Perakyla, 1998). At the same time the question of how well a staff handles this task has a high impact on how patients perceive their treatment (Pennbrant, 2013) and on the reputation of the organization. Learning how to conduct such conversations is not well supported in medical schools, and training practices such as role playing often fall short in real-life situations (Delvaux et al., 2005). In our ward, as well, this was also described as a problem, and we therefore found reflection on past conversations done together with others of the staff to be a promising means to support the physicians of the ward in their daily learning (Prilla et al., 2013, 2012).

3. THE TALKREFLECTION APP TO SUPPORT REFLECTION AT WORK

This chapter describes and reflects on a design process aiming at the support of bottom-up reflective learning for physicians. The tool developed for this support was called "TalkReflection App". While it is a result of the design process to be described in Section 4 and while our reflections in Section 5 are closely connected to its design and uptake, the tool is not at the center of this chapter. Therefore we provide a brief overview of the tool in this section, which eases the understanding of this chapter.

The TalkReflection App aims to support the different phases of reflection in health care (Maiden et al., 2013; Prilla, 2015; Prilla et al., 2012; Prilla and Renner, 2014). There is a need to support the documentation of experiences during work, to reflect on them individually, and to share them in order to be able to reflect collaboratively, for example, by providing similar experiences or arriving at changes for future behavior together from shared experiences. In addition, there is a need to share results with others, so that people who could not take part in the reflection process can still benefit from the outcomes. The TalkReflection App provides support for the following phases:

- **Capturing/documenting conversations**: The app supports documenting conversations on mobile and desktop devices. Users can thus write down, rate, share, and access their experiences whenever they want. If, for example, a physician in a hospital had a difficult conversation, she may write down the course of the talk, including what she thinks went wrong, and rate the conversation as bothering (see Fig. 9.1, #1).
- **Individual reflection**: By accessing experiences documented in the app and adding personal comments on them, users can reflect individually and sustain these reflections in the app. The physician from our example may add a comment on the documented experiences, which expresses that she thinks the relative was not well prepared for the message she had to convey to her (see Fig. 9.1, #2).

FIGURE 9.1 The TalkReflection App.

- **Collaborative reflection**: Users may share documented experiences, they can access experiences shared with them (see Fig. 9.1, #4), and they can leave comments on all experiences available to them (#3). In our example, a fellow physician could use such a comment to describe a similar situation and suggest a solution.
- **Sustaining outcomes**: If users come to what they think may be a solution for situations they reflected upon, they can write it down in the app (tab "Outcomes, see Fig. 9.1). In our example, the physician may note that colleagues should better inform a senior physician before potentially difficult conversations.

In addition to the process laid out below, the app was evaluated in multiple cases, among them the case described above. Results showed that it was used for reflection, but that there were differences in how it was applied in different contexts and by different professional groups (Prilla, 2015, 2014; Prilla et al., 2015; Prilla and Renner, 2014).

4. DESIGN PROCESS AND RESULTS

4.1 Methodology

The process of designing and evaluating the TalkReflection App described previously was run in a three-step approach using a variety of methods well known to socio-technical design, which will be described below. Physicians and other stakeholders on the ward were tightly integrated into every step (Table 1.1).

4.2 Step 1: Ethnography and Interviews to Explore the Domain

As described previously, there was not a sufficient body of information on how ICT might support reflection in health care practice at the time we began working on the project. Therefore in the initial step, we conducted exploratory, ethnographic research, including interviews and observations. We observed a physician and a nurse for 2 days each and interviewed four more people in the hospital (see Prilla et al., 2012; Prilla et al., 2013 for details) (Table 1.1).

TABLE 1.1 Steps Taken in the Socio-technical Design of the Cases Described in This Chapter

	Participants	Time	Purpose	Methods
Ethnographic study	1 physician, 1 nurse	2 days per person	Understand reflective practice and support needs	Field notes, interviews
Prototyping	3–6 physicians (varying over workshops)	4 half-day workshops	Progress from paper prototype to running system	Workshops, prototype walkthroughs
Formative evaluation	6 physicians	Half-day workshops, 4 weeks usage	Understand usage and integration into practice	Workshop, interviews, data analysis

We followed a nurse and a physician 2 days each for their entire shifts. Based on an analysis scheme, we noted down all the circumstances of their work, with a special focus on the occurrences of reflective interaction. The purpose of all observations was to understand the work done in health care environments and particularly to use the way people discuss and learn about their practices as a basis for tool development. Our analysis included situations and habits of communication and cooperation, constraints imposed by the workplace, and the actual practice of reflection. We then compared our observations to the small body of literature available. For this, we used an observation scheme containing aspects of reflection such as interaction with colleagues (participants, place, time, etc.), occurrences of reflection (participants, topic, data used, etc.), and technology used (purpose, relation to work, etc.). The resulting notes were transcribed and coded with the categories from the observation scheme.

Interviews were conducted with the observed workers and additional staff to clarify rationales, needs, and wishes of staff with respect to reflection. The interviews lasted up to 60 min and contained questions about the interviewees' workplace, its special characteristics, aspects of learning and motivation in daily work, communication and collaboration during the day, as well as existing and envisioned practice of individual and collaborative reflection. Interviews were audiotaped and transcribed literally and were then coded by two researchers (Prilla et al., 2013, 2012).

The field notes and interviews proved to be a good basis for tool development, and for discussing with the medical and management team whether the tool development would fit their needs in reality. There is, as we realized later on and were told by hospital staff during later design processes, also a need to involve management and other relevant stakeholders such as the board of employees to identify typical issues such as time and resource problems—awareness of these issues and how they constrain design is crucial in medical settings.

Our approach in step 1 revealed several issues to be dealt with in the communication and cooperation of staff. Among them, communication between relatives and physicians was often mentioned as emotionally distressing for physicians (and patients), but important for the treatment of patients and the reputation of the hospital. We were also able to witness some of these conversations and to observe the emotional stress that emerged on both sides (physician and relatives). As mentioned in Section 2, learning how to conduct these conversations is a well-known challenge among young physicians. However, there was no approach implemented to support this learning in the hospital we worked with. Physicians told us that they would very much appreciate better opportunities to share experiences from these conversations with colleagues and superiors in order to learn how to improve. We took this as a case to work on, and in what follows, we describe our efforts in developing a socio-technical solution to enable physicians to learn from each other.

4.3 Step 2: Participatory Design Workshops and Prototyping

The second step in the study was devoted to participatory design of prototypes that could be used by physicians. This step was done in multiple workshops and using different versions of prototypes. We started with a paper prototype (see Fig. 9.2) and used several stages of development of the later tool as prototypes. We had created a vision for the prototype

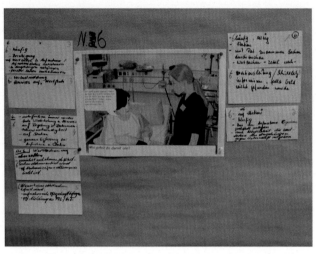

FIGURE 9.2   Results of the pretest of the process blueprint for collaborative reflection support.

in which we wanted a tool to present scenes from work experiences to practitioners and to enable these practitioners to comment on these scenes as a means of articulating their reflections (Prilla et al., 2012). The paper prototype in Fig. 9.2 represented work situations with an abstract picture in the center and allowed comments by writing on paper cards and attaching them around the scene described.

In the workshop on the paper prototype, five employees of the ward (four nurses and a physician) were asked to reflect on the scenes depicted on five poster walls similar to the example shown in Fig. 9.2. These scenes were taken from our observations and interviews on their daily work and represented, for example, a patient complaining or a patient missing her valuables. For each scene, we asked the participants to follow a scripted process, where we asked them to first look at the scenes, then to write down similar scenes they had experienced, then to share their experiences by pinning the cards to a board and explaining them to the others. Finally we asked them to discuss the different comments shared on each scene. This helped us to find out whether our vision was applicable in practice and whether it could provide a benefit.

Next, we conducted three workshop sessions: First, we ran a workshop with three physicians, who used an early prototype of the app that was aligned to the paper prototype. The physicians were asked to give feedback on its applicability and potential utility. Second, we conducted two workshop sessions with two physicians each. Each session followed a similar structure (Prilla et al., 2012): In the beginning we explained the tool (as there were new participants in each round) and the changes we had applied.

One of the integral and most helpful parts of the workshops was the hands-on session in each workshop, which enabled the participants to use the respective prototype. In the first workshop, we asked all participants to use the prototype simultaneously to enter experiences, to share them, and to comment on shared experiences (Fig. 9.3). After that, we discussed the usage experience in the group. In the other workshops, we asked small groups of participants (mostly dyads) to use the prototype in the same way, but in a sequence of separate sessions—this was done to accomplish focused work despite emergency calls or other urgent tasks. These hands-on sessions created interesting information in each workshop, including uncovering usability issues, potential barriers to using the prototype, and information on how to embed it into the work processes of the respective workplace. People made proposals for improving the early prototypes and also came up with proposals for organizational support of using the app for reflection. One physician, for example, told us that he and his colleagues might like the app because they could take it with them and use it during small breaks in their daily schedule (as opposed to doing all documentation at one time after work).

4.4 Step 3: Evaluation in Practice

After the prototyping phase, and being aware of the fact that in practice additional needs and constraints may come up, we conducted formative evaluation sessions. The main purpose of these was to get feedback from the staff on the usage of the app in their practice. We ran workshops before we gave the app to the physicians for a trial period (4 weeks) and after they used it. In addition, there was an interim workshop in the middle of this period (see Prilla, 2014, 2015). We gathered the participants in an initial workshop and showed them how to use the app for reflection, including hands-on usage of the app. For example, physicians were asked to look at shared reports and create comments. At the end of this workshop we discussed questions about the usage and opportunities to use the app during work. We then left it to the physicians how and when they would use the app.

After 2 and 4 weeks of using the app we ran workshops to get feedback on the usage and to simulate meetings in which the app was used for face-to-face reflection. Physicians were asked to choose one or two topics shared in the tool, to explain the topic to the others and to collaboratively reflect on it. After this, discussions were held as to whether and how this could become part of their practice. In the workshops after 4 weeks, we also conducted interviews with users in order to get their impression of using the app. After this formative evaluation we improved the app to better fit the workplace and then conducted a summative evaluation, investigating whether and how the app could help the participants to reflect more often and

FIGURE 9.3   A physician using the TalkReflection App on a tablet device during the second evaluation workshop.

get better results. Among other results (Prilla et al., 2015; Prilla and Renner, 2014), we found that the app was used by the physicians and that it was perceived as helpful for sharing experiences with others. We also found that there were some constraints and obstacles in using the app as part of the daily work, which leads us to reflections on the design process.

5. REFLECTIONS: SOCIO-TECHNICAL DESIGN CHALLENGES IN HEALTH CARE

Our work, as laid out above, opens up room for reflection on constraints, challenges, and opportunities for socio-technical design in health care environments. It points to some challenges that seem to be typical for health care contexts. This reflection revealed several challenges for socio-technical design, among which we will present the five most significant below: dealing with hierarchies, low willingness to employ computers for communicating and documenting personal experiences, low willingness for renewing infrastructure, and ill-structured procedures and routines. In the following sections we characterize each challenge, describe our design approach to dealing with it, compare this approach with previous work in the literature and with our experience with other cases, and finally discuss our conclusions for meeting the challenge using suitable approaches to socio-technical design.

5.1 Challenge 1: Dealing With Inherent Hierarchy Structures and the Role of Superiors

The importance of the often entrenched hierarchies in health care is well known and has been researched intensively; known issues such as power relations and the need for management "buy-in" are even more important here than in other environments. In the cases described previously, hierarchies also caused another, surprising effect. One senior physician wanted staff to be more active in learning at work. He was therefore considered a key person to use the tool in order to motivate assistant physicians to share and discuss experiences. However, he created an adverse effect using the app to provide advice rather than moderating reflection. As a result, once he had made a suggestion for how to deal with an issue, none of the assistants suggested different approaches or participated in further discussion. In a study run with a nursing home, we had another surprising effect, as the manager of the home stopped the usage of the app when she realized that she was not in perfect control of her staff's interactive learning (Prilla et al., 2015). This leads us to the notion that socio-technical design needs to deal with the particular difficulty of balancing change and inherent hierarchic structures.

Our **design approach** was that the usage of the tool by the different stakeholders involved was part of the discussion and hands-on in workshops, in which we agreed who would do what and when. We simulated meetings and moderation of reflection in the workshops. Senior personnel were asked how they would like to use the tool and encouraged to positively influence its usage.

With respect to **previous work and experiences**, the study by Walton (2006) states that the problem with exchanging experiences stems from the current focus on power, which in turn leads to the demand for a shift from hierarchically oriented learning toward guiding.

In similar contexts, product champions have been found to be helpful in this. More specifically, during a project on support for trauma teams in pediatric intensive care, Kusunoki et al. found problems related to different information needs between doctors and nurses caused by hierarchies (Kusunoki et al., 2014). They uncovered this problem in participatory design workshops and dealt with it by creating different visualizations that would match the needs of different roles.

In our own previous work we have also found hierarchies blocking the exchange of experience and opinions regarding, for example, organizational procedures. For instance in a previous case of a training company, people were not willing to invest time to understand how procedures took place, they did not look at models of the processes, and mainly tried to guess the position of their superior. We realized that there was a need to ensure that everybody's voice would be heard, offering anonymity to contributors. Furthermore, people's contributions were noted down and displayed on a board to keep them available throughout the whole discussion. This case (cf. Herrmann et al., 2002) initiated the idea of organizing and facilitating an organizational walkthrough (Herrmann et al., 2002; later "socio-technical walkthrough": Herrmann, 2009) in which all participants could make a valuable contribution based on their experience and in which the value of this experience does not depend on the person's hierarchical position. The workshop was guided by a set of questions that were iteratively applied to details of an artifact or document and every trainer involved in the project was asked their opinion. In the TalkReflection case, it might have been worth a try to run structured, facilitated communication in a meeting at the beginning of the trial, which would have encouraged the notion that with respect to certain questions (for instance how to speak with relatives) everybody can be an expert because of their experience or communicative competences.

When **reflecting with hospital staff** about how to deal with this issue, it was suggested to train the superiors in health care (e.g., senior physicians) to facilitate bottom-up learning in order to make it happen and the staff in general on how they can benefit from learning together. We integrated this in our workshops to create awareness of the need for facilitation and to strengthen self-efficacy for learning, thus supporting bottom-up processes in health care.

In sum, hierarchies can be considered as a typical issue for socio-technical design in health care. This does not only concern a potential conflict between a strict hierarchy and self-regulated, bottom-up processes. Rather than that, there is often a need for slight and partial shifts in roles and responsibilities of superiors, such as becoming a guide and facilitator in reflection on conversations rather than an all-knowing superior, and leaving the creation of solutions for problems in daily work to staff rather than providing staff with solutions.

In our case, the challenge was to integrate bottom-up support for learning from conversations that would be used mainly by young physicians, and in which superiors would help them to learn. The need for this integration is supported by the previously mentioned literature and fits the needs communicated during our work with the physicians. Therefore we may regard socio-technical design as implementing change, such as (in our case) bottom-up, self-directed processes, into a hierarchical, top-down structure that is inherent to and important for health care. Then, design approaches such as structured walkthroughs and visualizations can create awareness and strengthen self-efficacy, thus supporting bottom-up processes in health care.

We found that we were not able to create this shift: The challenge in our case was whether there would be a chance to slightly and partially change a social structure in which a power hierarchy has good reasons for existing, such as responsibility for the well-being of patients and residents and is thus likely to remain in place. Therefore, the question is whether and to what extent socio-technical design can push the boundaries of hierarchical structures to design bottom-up processes in some areas of the work of health care professionals while retaining the positive effects of hierarchy. However that may be, there is a clear need to train superiors in health care to facilitate bottom-up learning in order to make it happen.

5.2 Challenge 2: Low Willingness to Use New Technology for Communication Purposes; Low Expectation of Benefits From Employing the New Technology

At the health care workplaces studied, we could not find a culture of using technology for collaborative learning as initially intended or for intensive communication support. By contrast, employing new technologies was primarily focused on diagnosis and therapy. Secondarily, IT is used to support administrative documentation (Bossen and Jensen, 2014). In health care settings, face-to-face communication among peers is a well-established culture and employed by preference for mutual support in emotionally stressful situations. Computer-mediated communication is perceived as possibly interfering in this context. Similarly, the exchange of experiences being a possible basis for reflection happens mostly face-to-face. The availability of the experiences was consequently limited: In the beginning, there was reluctance in exchanging experiences with the TalkReflection App. They claimed that they did not have enough time to use the TalkReflection App. However, time pressure seems not to be the main acceptance barrier since some employees found enough time after they realized the benefit of the tool.

The **design approach** to meet this challenge was based on participatory meetings, where possible usage scenarios were presented, repeatedly discussed, and trialed in the workshop by role play. Superiors were tightly integrated into the planning of the tool usage. As a crucial result, the content collected and conveyed with the TalkReflection App became topics of the regular staff meetings. This organizational measure helped to demonstrate the potential benefit of the TalkReflection App. Such a socio-technical process appeared more promising than improving the provided tools or offering a variety of technical methods (such as digital pens, tablets, etc.) to simplify documentation tasks and the requested feedback from participants. Besides the participatory meeting, interviews were helpful to understand the communication culture and associated worries. For instance, it turned out that the potential authors of experience reports felt that these would not be new to the potential readers.

From **previous work and experiences**, we knew that people might refuse to learn to employ new technologies for well-established tasks if the expectable benefits are not substantial and clear. Therefore we usually helped people to understand and to imagine the usefulness of the new technology by simulating its use with playful scenarios during workshops (Loser and Herrmann, 2002). This is especially important in the case of supporting asynchronous collaboration, where the benefit of certain contributions is not immediately recognizable for the users. We tried to help participants to understand their daily processes and routines step-by-step and to identify how the new technology could be most beneficially integrated. It is reasonable that the potential users very deliberately plan and discuss their collaboration and coordination (Carell et al., 2005) or at least they should be extensively informed about the organizational procedures into which the new technology should be embedded (Kienle and Herrmann, 2004). It turned out to be important that people can easily be aware of what will happen with their documented problems or proposed solutions. New technologies are usually accompanied by organizational change (Baxter and Sommerville, 2011), for example, the documentation of experiences or problems can shift from an individual to a collaborative task (Kienle and Herrmann, 2004). Support of existing tasks and needs may be considered as guidance for introducing new technology (Eason, 1988). However, in the course of organizational change, new tasks may develop which are hard to anticipate. Referring to the study by Cherns (1987), the challenge is to deal with incompleteness since the change process cannot be described as a transition from one stable state to another, but between one period of transition to another.

When **reflecting on this issue with hospital staff**, they felt that the highest acceptance barriers were observed if new tasks or goals (such as intensifying learning by collaborative reflection) were combined with a new technology to which staff were not accustomed (like the TalkReflection App). For such constellations they suggested adopting a lead user approach (e.g., using an experienced and respected member of staff as a lead user) and to start with supporting tasks technically that users are already familiar with.

The **discussion** of the low willingness to employ the TalkReflection App needs to reflect a certain role of new technology: The TalkReflection App was intended to increase the employees' effort at reflecting on their work. The new technology served more as a trigger for carrying out new tasks than as a support for existing tasks. In this case, *technology* can be a seed that supports evolutionary growth (Fischer and Ostwald, 2002) of an organization's way of establishing new types of learning. Within such constellations, socio-technical design has to rely on the effect of incremental steps. One example could be to identify lead users who can serve as a role model for facilitating perspective sharing (avoiding the effects of hierarchical dominance of opinions). Another strategy is to identify tasks that are already accepted by the potential users—such as participating in a locally distributed community—and which promote understanding of the benefits of the technology.

5.3 Challenge 3: Dealing With Little Support for Appropriate Technical Infrastructure, Especially for Nonmedical Tasks

In the hospital, we worked with (and in many other hospitals and health care sites) the level of ICT support is behind other industries (e.g., with respect to WiFi). One reason was the aforementioned culture that important tasks are carried out face-to-face with colleagues, patients, residents, and other parties, and technology is used for diagnosis, therapy, and medical documentation. Furthermore, IT management in the hospital was neither willing to nor capable of implementing technological upgrades and refused to allow us to replace or even extend the software infrastructure, mainly for security reasons. It was also not possible to interface with the available hospital information system due to licensing issues. This made it hard to run trials and to demonstrate the benefits of the

TalkReflection tool and to integrate it properly into the existing context of work. As a result and despite the benefits provided by using the app, this may have hindered larger impact of our solution.

As a **design approach** with respect to this challenge, we discussed use cases in workshops early on with different stakeholders and showed them how the goals could be reached with different kinds of devices. We tested prototypes of paper-based note taking and digital pens to see if this would come in handy, and we tried tablets in different sizes to make sure they could be carried around like notebooks in the physicians' white coats. Access to the internet, however, remained a problem due to a lack of WiFi access. Therefore we allowed the employees to enter data on tablets without connection to a network and manually synchronized it afterward, when there was time to move to a space in which network access was available. This, however, impeded timely interaction between the physicians. Ultimately the strategy was to bring the necessary equipment to the hospital by ourselves. First we brought tablet devices and finally we ended up with setting up a dedicated server within the hospital and made the tool available on computers on the intranet, which were available in the physicians' room. This solution took a long time for negotiation and was allowed us temporarily and for research purposes only.

We knew this infrastructure problem already from **previous work and experiences**, and it causes specific difficulties in early phases such as field testing for formative or summative evaluation. Consequently, we felt compelled to bring not only our own devices but also to offer infrastructure. For example, in a case where we introduced a continuous micro survey for measuring companies' creativity climate (Herrmann et al., 2011), we first ran the tool from our own Web server to store the anonymized answers. Only after this first try were some companies willing to roll out a desktop version with which it was much easier to display reminders to people and which asked them to answer some questions. The lesson learned is that socio-technical design needs phases where the usefulness and the proper functioning of a new technology can be demonstrated before one can expect a willingness to change the existing infrastructure. This is in accordance with other findings: Bossen and Jensen state that digitalization in medical work is well behind other workplaces due to constraints such as the aforementioned necessity for mobility (Bossen and Jensen, 2014). They argue that developing technology around existing purposes and technical support such as electronic patient records might be a good way to start, as these are central artifacts not only for documentation but also for the whole process of patient care.

During our **reflection on this challenge with hospital staff**, it was suggested to accompany socio-technical design in health care early on with a portfolio of offerings and requirements to implement them in order to overcome structural problems with IT infrastructure. Staff suggested that this should include a means to use devices without their network functions, to host tools and services offered to organizations on the servers of health care providers, and ways to abstain from using technology where it is not necessary.

A concluding **discussion** of the infrastructure challenge points toward the following dilemma: To make the users and their management aware of the benefits, a sufficient infrastructure is needed, for example, to exchange and reflect experiences in every situation and under the condition of mobility. However, before the management improves the infrastructure, they want to be convinced of the benefits that can be expected. This dilemma can only be solved if we demonstrate the benefits with low-tech prototypes or if we find ways to use an infrastructure that is independent from the site where a solution is evaluated, such as providing our own infrastructure or using laboratories, as suggested by (Houben et al., 2015), or (mobile) demonstrators.

5.4 Challenge 4: Dealing With Unstructured, Spontaneous Processes Driven by External Factors Versus Meaningful and Planned Integration

From observations during our studies and from the literature, we soon recognized that the work in health care workplaces is highly unstructured and hard to anticipate (Kusunoki et al., 2014; Prilla et al., 2012). Visiting and treating patients is affected by the everyday changing needs of the individual, as well as by external factors and contingencies such as when emergency patients arrive at the hospital. Assistant physicians find it hard to describe a typical workday in a sequential way. Therefore it proved difficult to identify the appropriate occasions where extra effort should be invested for reflective learning.

We tried a two-fold strategy as a **design approach** to deal with the unstructured procedures. First, we tried to anchor the tool usage to recurring events, such as team meetings, to create a regularity and rhythm of tool usage. Second, we designed the TalkReflection App in a way that allowed for short-time, asynchronous interactions (using the tool whenever there was time, spontaneously, see the study by Prilla et al., 2012). The users would then be able to discover that a session of using the app will not require their attention for a longer time period.

With respect to **previous work and experiences**, we found that people imply or maintain (e.g., within a department of a University) that well-anticipated and well-coordinated processes do not take place. Even if there might be resources for extra activities, it is unlikely that people will reveal this. In addition, these resources cannot be detected officially and explicitly (*cf.* Suchman, 1995 about the visibility of work practices). In another project, staff of a facility department referred to the emerging everyday challenges, which always arise suddenly. This seems to be typical for all cases where one has to deal with the behavior of other people, with spontaneous needs, or with emerging events—as in the medical area. To address this, we ran participatory workshops with the staff to analyze challenging events and to understand whether they take place regularly or whether they really occur exceptionally. This was a hard challenge since people were proud of their ability to react flexibly and knowledgeably to unforeseen challenges. They did not necessarily want to disclose that they have already developed routines for dealing with them. We tried to identify the more regularly occurring procedures and to analyze them to understand at which points in time occasions for coordinative work remain, such as documenting a case, taking notes, informing others, etc. Another obstacle is that it is nearly impossible to run a series of facilitated workshops with all relevant stakeholders since they always have reasons to stay away due to emergencies. Similarly, in the TalkReflection case the doctors were continuously on standby during the workshop and had to leave or at least answer phone calls if necessary. These limitations of analyzing and structuring work processes have been intensively discussed in the literature (Engeström, 2000; Schmidt, 1997; Suchman, 1995).

When **reflecting about his challenge with hospital staff**, it was suggested to plan extra personnel for socio-technical development. This should be coordinated in a way that the group that is available during those shifts when the researchers and designers are present should also be able to be present for workshops and interviews most of the time.

The **discussion** of this challenge can benefit from the perspective of meta-design, especially the concept of underspecification. This has been proposed as an approach that gives examples and a framework within which organizational procedures and the use of new technology can develop (Fischer and Herrmann, 2011). The meta-design perspective is also discussed in the health care context by pursuing the approach where socio-technical systems are built, which design themselves (Coiera, 2007). Outlining a rough scaffold of how the TalkReflection App could be embedded into daily routines is an example of meta-design. The employees can then find their own way in deciding how to use the TalkReflection App and to what extent. Meta-design would also be accompanied by meta-reflection about when and under which conditions reflecting on one's own experiences leads to mutual benefit.

5.5 Challenge 5: Dealing With Privacy, Liability, and Security With Respect to Data Handling

Handling patient health data is very sensitive and is specifically regulated by law in the European Union. When new software is introduced, organizations have to be especially careful about which data are collected, stored, and transferred for which purpose. On the one hand, reflection needs the documentation of experiences, especially those which are related to problems and erroneous actions. On the other hand, data about individual patients, problems, and errors could be a basis for challenging the hospitals liability for complying with privacy laws or for suing the hospital by patients or their relatives who demand compensation for supposed improper treatment. Another issue is that the documentation of problems is also available to the practitioners' colleagues in collaboration settings. The privacy of users who document and share experiences for reflection is a crucial aspect for the adoption of the technology. The study by Paterson (1995) stressed the importance of trust for collaborative reflection, as participants will not be willing to describe experiences where they feel they might have acted wrongly if there is not trust in the rest of the group.

Our strategy was to follow a privacy-by-design pattern right from the beginning. This included discussions with data protection officials and implementation of security measurements like encrypted servers. On the organizational level, users were instructed to describe cases without referring to individuals' names, to use pseudonyms, and not to describe medical conditions since these could have been used to identify people. These instructions were also conveyed with the app itself. Especially the highly educated staff in the hospital had a high awareness of the impact that disclosure of this data might have on patients' privacy. For others, this awareness was easily raised. Therefore, pseudonymity in the written stories was reached mainly by organizational means, but also with prompts and hints linked within the user interface. In addition, we had to deal with the privacy of the users, who were not always willing to share their problematic experiences with colleagues, especially when they felt it was their fault. To deal with this, we allowed anonymous postings and comments to be anonymous. Although there were fears that this might be abused, for example, cybermobbing, users remained respectful and professional in their postings. The whole design pattern is described in the study by Degeling and Nierhoff, 2013.

Our strategy was inspired by **earlier projects** where personal data were handled. We usually try to apply general privacy principles (Cavoukian, 2009) or more specific principles such as minimization of personal data and transparency about what is stored and processed (Gürses et al., 2011; Rost and Bock, 2011). However, we had to realize that the extent of applying these principles is decisively influenced by the culture of the organization where a socio-technical system is introduced. In one case in the chemical industry (Kienle and Herrmann, 2004), the management and employees suggested it would be inappropriate to allow anonymous posting on a collaborative platform. In another project (Herrmann et al., 2011), the anonymity of the participants was a key selling point, although it caused an awkward handling of registration procedures and of providing remainders for participating in the survey.

Regarding liability and the appropriate dealing with privacy issues, we have often introduced a facilitator (Kienle and Herrmann, 2004) to accompany the information exchange on a platform, especially when it is possible to post arbitrary text, for example, in a forum. Alternatively, a more technical and rigid approach would have been to minimize the extent of data—for example, by using restricted forms—in a way that there is no possibility to enter arbitrary text.

The discussion of this challenge has to conclude that in socio-technical design it is inevitable to make trade-offs between freedom of use and control of user input to preserve privacy. Privacy by socio-technical design means to check the appropriateness of privacy-by-design principles in each step of development. Therefore, managers, the designers, as well as software developers have to be aware of the trade-offs to be made and discuss them with all stakeholders of the design process. Especially when developing support in health care, it is necessary that users are aware of the privacy of third parties like patients and clients. It is advantageous if participants are either already aware of privacy issues or can be easily trained to adopt their behavior—for example, when taking notes—in a way that avoids collection of data about third parties. To achieve a sufficient level of privacy and trust is a matter of combining organization measures, training, and technical features and can be considered as "privacy by socio-technical design".

6. CONCLUSION

Specific socio-technical challenges emerge if a new type of technology such as the TalkReflection App is intended to trigger change (in this case to intensify collaborative reflective learning) and is intertwined with new tasks or goals. The starting point, in particular, is characterized by a dilemma: People and organizations are not willing to adapt their behavior, routines, or infrastructure, before the benefits of the change are demonstrated. However, these benefits can only become apparent for technologies that support collaborative tasks if a sufficient number of participants are willing to test it and if a sufficient infrastructure that supports the testing is available. Subsequently the phase of formative evaluation throughout socio-technical design has to be based on surrogates of networking infrastructure or with the help of low-tech prototypes. In the health care sector, this dilemma is intensified by several factors:

- Technological support is focused on medical tasks and administrative documentation. The medical staff are used to face-to-face communication, especially for exchanging experiences and for supporting mutual learning. Adopting new technologies together with new types of behavior (e.g., intensifying reflection-based learning) requires the identification of lead users who are ready to try out the new tools and to demonstrate their benefit and who are willing to adapt their habits and role patterns.
- Learning and adaptation of behavior takes place under the influence of superiors within a strict hierarchy. Introducing technologies that support and generate benefits from a bottom-up approach has to be aligned with organizational and cultural change. Agile approaches including trial-and-error and testing of incompletely specified tools and procedures are faced with the stricter conditions of hierarchy-oriented routines and regulations. To push the boundaries or organizational structures by socio-technical design requires the facilitation and training of superiors.
- Everyday routines and processes are continuously interrupted by emergencies, and each case and patient has specific characteristics to be dealt with. As a result, carrying out workshops for participatory design is challenged by non-anticipatable workflows and interrupting emergencies. Organizational change and the usage of new technology cannot be seamlessly achieved by implementing predefined workflows and routines. By contrast, scaffolds have to be offered, which help lead users to find the most appropriate ways to integrate the new technology into their daily routines.
- Since data about the condition of patients and their treatment are considered quite sensitive, a trade-off between freedom of use versus control of user input and data transfer has to be taken into account. Users and managers have to be made knowledgeable about privacy principles and about ways to avoid the collection of data about individuals or third parties.

Given these conditions, detailed planning of organizational changes cannot take place in advance but is rather subject to an evolutionary process. This process includes a fluent transition between the following:

1. The phases of evaluation-based improvement and regular usage.
2. The participatory design, which takes place before use, and design-in-use (*cf.* Fischer and Herrmann, 2011), where the tool is already adopted within daily routines but is still a subject of adaptation to individual needs and the characteristics of tasks.
3. Improvised technological support with prototypes or preliminary infrastructure and systematically adapted and well-approved infrastructure.

All in all, in the context of our experience, it appears to be a necessity that technology that is not tightly connected with medical treatment but supports communication and coordination or learning issues, needs a deliberate and flexible phase of organizational and cultural change that supports the development of acceptance and new routines.

Acknowledgments

This work has been supported by the European commission (project MIRROR, project number 257617). The authors thank all members of the project for their support and ideas on this work. In particular, Dominik Walter has contributed his insights as a health care practitioner to the chapter, and Martin Degeling has contributed to the discussion of the challenges associated to privacy. They thank both for their valuable input.

References

Baxter, G., Sommerville, I., 2011. Socio-technical systems: from design methods to systems engineering. Interacting with Computers 23, 4–17.

Bossen, C., Jensen, L.G., 2014. How physicians 'achieve overview': a case-based study in a hospital ward. In: Proceedings of the 17th ACM Conference on Computer Supported Cooperative Work & Social Computing. ACM, pp. 257–268.

Boud, D., 1985. Reflection: Turning Experience into Learning. Kogan Page, London.

Carell, A., Herrmann, T., Kienle, A., Menold, N., 2005. Improving the coordination of collaborative learning with process models. In: Proceedings of CSCL 2005. Computer Supported Collaborative Learning 2005: The Next 10 Years. Lawrence Erlbaum Associates, pp. 18–27.

Cavoukian, A., 2009. Privacy by Design – the 7 Foundational Principles.

Cherns, A., 1987. Principles of sociotechnical design revisted. Human Relations 40, 153–162.

Coiera, E., 2007. Putting the technical back into socio-technical systems research. International Journal of Medical Informatics 76, S98–S103.

Degeling, M., Nierhoff, J., 2013. Privacy-by-design am Beispiel einer Anwendung zur Unterstützung kollaborativer Reflexion am Arbeitsplatz. In: Proceedings INFORMATIK 2013, Lecture Notes in Informatics (LNI). Presented at the INFORMATIK 2013-Workshop "Der Mensch im Fokus: Möglichkeiten der Selbstkontrolle von Datenschutz und Datensicherheit durch den Anwender,". Köllen Druck+Verlag GmbH, Koblenz, pp. 2060–2071.

Delvaux, N., Merckaert, I., Marchal, S., Libert, Y., Conradt, S., Boniver, J., Etienne, A.-M., Fontaine, O., Janne, P., Klastersky, J., Mélot, C., Reynaert, C., Scalliet, P., Slachmuylder, J.-L., Razavi, D., 2005. Physicians' communication with a cancer patient and a relative. Cancer 103, 2397–2411. http://dx.doi.org/10.1002/cncr.21093.

Eason, K., 1988. Information Technology and Organisational Change. Taylor and Francis, London.

Engeström, Y., 2000. From Individual Action to Collective Activity and Back: Developmental Work Research as an Interventionist Methodology. Workplace studies, pp. 150–166.

Fischer, G., Herrmann, T., 2011. Socio-technical systems: a meta-design perspective. International Journal of Sociotechnology and Knowledge Development (IJSKD) 3, 1–33.

Fischer, G., Ostwald, J., 2002. Seeding, evolutionary growth, and reseeding: enriching participatory design with informed participation. In: Proceedings of the Participatory Design Conference. PDC, pp. 135–143.

Forneris, S.G., Peden-McAlpine, C.J., 2006. Contextual learning: a reflective learning intervention for nursing education. International Journal of Nursing Education Scholarship 3. http://dx.doi.org/10.2202/1548-923X.1254.

Gürses, F.S., Troncoso, C., Diaz, C., 2011. Engineering privacy by design. Computers, Privacy & Data Protection.

Herrmann, T., 2009. Systems design with the socio-technical walkthrough. In: Whitworth, B., de Moor, A. (Eds.), Handbook of Research on Socio-Technical Design and Social Networking Systems. Information Science Reference.

Herrmann, T., Carell, A., Nierhoff, J., 2011. Creativity barometer: an approach for continuing micro surveys to explore the dynamics of organization's creativity climates. In: Proceedings of the 8th ACM Conference on Creativity and Cognition, C&C '11. ACM, New York, NY, USA, pp. 345–346. http://dx.doi.org/10.1145/2069618.2069688.

Herrmann, T., Hoffmann, M., Kunau, G., Loser, K.-U., 2002. Modeling cooperative work: chances and risks of structuring. In: Cooperative System Design. A Challenge for the Mobility Age (Coop 2002). IOS Press, pp. 53–70.

Houben, S., Frost, M., Bardram, J.E., 2015. Collaborative affordances of hybrid patient record technologies in medical work. In: Proceedings of the 18th ACM Conference on Computer Supported Cooperative Work & Social Computing. ACM, pp. 785–797.

Kienle, A., Herrmann, T., 2004. Collaborative learning at the workplace by technical support of communication and negotiation. In: Adelsberger, H.H., Rombach, H.D., Eicker, S., Pohl, K., Wulf, V., Krcmar, H., Pawlowski, J.M. (Eds.), Multikonferenz Wirtschaftsinformatik (MKWI). Presented at the Multikonferenz Wirtschaftsinformatik (MKWI). Akademische Verlagsgesellschaft, Berlin, pp. 43–57.

Kusunoki, D.S., Sarcevic, A., Weibel, N., Marsic, I., Zhang, Z., Tuveson, G., Burd, R.S., 2014. Balancing design tensions: iterative display design to support ad hoc and multidisciplinary medical teamwork. In: Proceedings of the SIGCHI Conference on Human Factors in Computing Systems. ACM, pp. 3777–3786.

Loser, K.-U., Herrmann, T., 2002. Enabling factors for participatory design of socio-technical systems with diagrams. In: Binder, T., Gregory, J., Wagner, I. (Eds.), PDC 02 - Proceedings of the Participatory Design Conference. CPSR, Palo Alto, CA, pp. 114–143.

Maiden, N., D'Souza, S., Jones, S., Müller, L., Pannese, L., Pitts, K., Prilla, M., Pudney, K., Rose, M., Turner, I., Zachos, K., 2013. Computing technologies for reflective and creative care for people with dementia. Communications of the ACM 56, 60–67.

Mann, K., Gordon, J., MacLeod, A., 2009. Reflection and reflective practice in health professions education: a systematic review. Advances in Health Sciences Education 14, 595–621.

Maynard, D.W., 2003. Bad News, Good News: Conversational Order in Everyday Talk and Clinical Settings. University of Chicago Press.

Paterson, B.L., 1995. Developing and maintaining reflection in clinical journals. Nurse Education Today 15, 211–220. http://dx.doi.org/10.1016/S0260-6917(95)80108-1.

Pennbrant, S., 2013. A trustful relationship—the importance for relatives to actively participate in the meeting with the physician. International Journal of Qualitative Studies on Health and Well-being 8.

Perakyla, A., 1998. Authority and Accountability: The Delivery of Diagnosis in Primary Health Care. Social Psychology Quarterly, pp. 301–320.

Prilla, M., 2015. Supporting collaborative reflection at work: a socio-technical analysis. AIS Transactions of Human-computer Interaction 1–16.

Prilla, M., 2014. User and group behavior in computer support for collaborative reflection in practice: an explorative data analysis. In: Rossitto, C., Ciolfi, L., Martin, D., Conein, B. (Eds.), COOP 2014-Proceedings of the 11th International Conference on the Design of Cooperative Systems. Springer.

Prilla, M., Degeling, M., Herrmann, T., 2012. Collaborative reflection at work: supporting informal learning at a healthcare workplace. In: Proceedings of the ACM International Conference on Supporting Group Work (GROUP 2012), pp. 55–64.

Prilla, M., Nolte, A., Blunk, O., Liedtke, D., Renner, B., 2015. Analyzing collaborative reflection support: a content analysis approach. In: Proceedings of the European Conference on Computer Supported Cooperative Work (ECSCW 2015).

Prilla, M., Pammer, V., Krogstie, B., 2013. Fostering collaborative redesign of work practice: challenges for tools supporting reflection at work. In: Proceedings of the European Conference on Computer Supported Cooperative Work (ECSCW 2013).

Prilla, M., Renner, B., 2014. Supporting collaborative reflection at work: a comparative case analysis. In: Proceedings of ACM Conference on Group Work (GROUP 2014). ACM.

Reynolds, M., 1999. Critical reflection and management education: rehabilitating less hierarchical approaches. Journal of Management Education 23, 537–553.

Rost, M., Bock, K., 2011. Privacy by design and the new protection goals. European Privacy Seal.

Schmidt, K., 1997. Of maps and scripts —the status of formal constructs in cooperative work. In: Proceedings of the International ACM SIGGROUP Conference on Supporting Group Work: the Integration Challenge: the Integration Challenge, pp. 138–147.

Schön, D.A., 1983. The Reflective Practitioner. Basic books, New York.

Suchman, L., 1995. Making work visible. Communications of the ACM 38.

Teekman, B., 2000. Exploring reflective thinking in nursing practice. Journal of Advanced Nursing 31, 1125–1135.

Walton, M.M., 2006. Hierarchies: the Berlin Wall of patient safety. Quality and Safety in Health Care 15, 229–230.

Double-Loop Health Technology: Enabling Socio-technical Design of Personal Health Technology in Clinical Practice

Jakob E. Bardram[1], Mads M. Frost[2]

[1]Technical University of Denmark, Lyngby, Denmark; [2]IT University of Copenhagen, Copenhagen, Denmark

1. INTRODUCTION

Due to demographic changes and poor lifestyle behavior, the modern healthcare systems are facing a significant increase in noncommunicable diseases. Such diseases are chronic or of long duration, with no or limited cure, and generally progress slowly requiring continuous care and treatment. The main types of noncommunicable diseases are cardiovascular diseases (like heart attacks and stroke), cancers, chronic respiratory diseases (such as chronic obstructive pulmonary disease (COPD) and asthma), diabetes, and mental health problems (like depression and schizophrenia). According to WHO, noncommunicable diseases are responsible for 63% of all deaths worldwide (36 million out 57 million global deaths) (World Health Organization, 2014). However, noncommunicable diseases are preventable through effective interventions that tackle shared risk factors, primarily tobacco use, unhealthy diet, physical inactivity, harmful use of alcohol, and stressful lives. Motivated by these challenges, a new class of healthcare systems target at personal and mobile health management is being researched and designed. This approach—called "Personal Health Technology"—is aimed at designing embedded sensor systems and using mobile and wearable computers for a novel pervasive, user-centered, and preventive healthcare model (Arnrich et al., 2010; Bardram, 2008; Bardram and Frost, 2016). The overall design paradigm behind most of these applications is to manually and automatically sample data from smartphones and wearable sensors, thereby enabling users to monitor important health and behavioral parameters, visualize these parameters, keeping users informed about their physical and mental state, reminding them to perform specific tasks, providing feedback on the effectiveness of their behavior, and recommending healthier behavior or actions (Consolvo et al., 2014).

Most of these personal health technology applications have focused on what we in this chapter will label as "single-loop" treatment. A single-loop application is focused mainly on patient self-management of well-being, healthy living, and/or disease care and treatment. However, in this chapter we will focus on what we will call "double-loop"[1] personal health technology in which both the patient as well as the clinician in the greater healthcare system are in the loop. As such, focus is on exploring how these upcoming personal health technologies may fit with, and integrate into, the existing healthcare system. In order to embed and utilize personal health technology as part of the wider healthcare system, a wide range of issues and challenges arise. On the one hand, such technologies offer the opportunity to improve quality of the existing treatment as well as doing it in a more efficient manner. For example, continuous monitoring of blood glucose, diet, and physical exercise is highly valuable in the treatment of diabetic patients and a personal health application that collects and manage such data would qualitative improve the clinical treatment of diabetes (Cafazzo et al., 2012). On the other hand, the introduction of such technologies into the existing organization and practice of an established healthcare system may pose significant challenges, which has not even been considered yet. For example, a recent survey of 656 personal health technologies for diabetes management showed that only 3.7% were designed for both patients and physicians—the vast majority (96%) was designed solely for patient self-management of diabetes (Arnhold et al., 2014).

The focus of this chapter is to present and discuss this "double-loop" treatment setup for personal health technology, or—in other words—to investigate the socio-technical setup that is needed for such emerging technologies to become an embedded part of the established healthcare system. This is a classic example of a socio-technical design, which seeks to codesign both the technological platform as well as the clinical organization around it but applied in a new setting of these emerging personal health technologies. A core tenet in the chapter is that a "single-loop" system may not take into consideration the larger socio-organizational context of these personal health technologies and therefore may encounter significant challenges if, or when, they are used in a professional—rather than personal—setup.

The chapter is based on a case of designing a personal health technology application for mental health, especially focusing on affective mood disorders (depression and bipolar disorder), called the MONARCA[2] system. The MONARCA system was designed to be a personal

---

[1] The use of the term "double-loop" in this chapter is not related to the concept of "double-loop learning", normally used in the context of organizational learning. There is no organizational learning involved in the present double-loop treatment model; it is merely an illustration of how two loops—the patient and the clinical loop—are intertwined in a socio-technical treatment setup.

[2] The MONARCA system has been developed over three main iterations. The initial design and technical implementation of version 1.0 is documented in Bardram et al. (2012) and a thorough usability evaluation is documented in Bardram et al. (2013). Version 2.0 incorporated a predictive feature as documented in Frost et al. (2013). Finally, the system is now available as a commercial product called the "Monsenso" system (see http://www.monsenso.com/). In this chapter, we are referring to version 1.0 of the system, which is the version that has been subject to most extensive clinical evaluation.

health technology for self-management of mood disorders for the patient, while also being integrated into the clinical treatment managed by the overall healthcare system (e.g., a hospital or a community mental healthcare clinic). In the chapter, we introduce the MONARCA case and describe the user-centered codesign of the system, the system design, its pilot testing, and its subsequent use in clinical trials. The MONARCA case provides a substantial empirical case due to its length and the number of people involved in the design and use of the system; the initial design of the MONARCA system was initiated in the spring of 2010; it has been subject to three major revisions; it has involved 20+ users (patients, psychiatrists, psychologists, nurses, designers, and IT specialists) in the design; it has been subject to five clinical trials; and it has been deployed and hosted as part of the IT infrastructure of a Danish Health Management Organization (HMO). The chapter will report on the lessons learned from the MONARCA case. These lessons relate to the *use* of the system (who uses it, when, and for what?), *organizational change* as part of socio-technical codesign (how can clinical treatment in mental health be changed to utilize such new technologies?), *scalability* (what are the implication for scaling such a system up?), and *socio-technical integration* (how does such a system integrate with existing systems and the workflow around them?). Based on these lessons, the chapter concludes with a set of recommendation of how to set up and utilize personal health technology in a double-loop manner in a clinical setting.

2. BACKGROUND: PERSONAL HEALTH TECHNOLOGY

The world is facing a health crisis. An aging population combined with physical inactivity, poor diet, and other poor lifestyle behaviors (e.g., stress and insufficient sleep) are contributing to an epidemic of chronic conditions, including obesity, diabetes, cardiovascular disease, and mental health problems (incl. stress, depression, and anxiety). In 1990 chronic and lifestyle-related diseases accounted for 66% of all healthcare expenditure in the United States (Hoffman et al., 1996)—in 2010 this had increased to 86% (Gerteis et al., 2014)—and worldwide; chronic diseases are responsible for 63% of all deaths (World Health Organization, 2014). At the same time, in the contemporary healthcare model, patients are diagnosed in late stage of a (chronic) disease and receive hospital-based treatment until they are discharged and no longer treated by the hospital. This "fix-and-forget" model fits very poorly with the challenges in chronic diseases, which to a much larger degree calls for a proactive, preventive, and continuous treatment model.

Motivated by these challenges a number of personal health technologies targeting a wide number of such chronic diseases and lifestyle problems have been introduced (Bardram and Frost, 2016). The core goal is to create technologies that are personal, participatory, preventive, and predictive and thereby engage patients in their own health. The overall design paradigm behind most of these applications is to manually and automatically sample data from sensors and smartphones and use this to provide patients with an awareness of their illness and give recommendation for treatment, care, and healthy living.

Personal health technologies can be grouped into three broad categories. The first set of systems can be labeled "wellness" applications, which seeks to "persuade" users into healthy behavior change such as encouraging physical activity (Lin et al., 2006; Consolvo et al., 2008), healthy eating habits (Pollak et al., 2010), or better sleep (Bauer et al., 2012). Lately, systems like the BeWell application have proposed a more comprehensive smartphone-based approach that can track activities that impact physical, social, and mental well-being namely, sleep, physical activity, and social interactions and provides intelligent feedback to promote better health (Lane et al., 2011). The second category enlist systems targeting management of chronic somatic diseases like diabetes (Mamykina et al., 2006; Smith et al., 2007), chronic kidney disease (Siek et al., 2006), and asthma (Lee et al., 2010). The third category includes systems targeted at mental health and illness, such as stress (Ferreira et al., 2008), depression (Robertson et al., 2005; Nakonezny et al., 2010; Burns et al., 2011), and more general purpose mobile phone systems for mood charting to be used in cognitive behavioral therapy (CBT) (Bauer et al., 2006; Matthews and Doherty, 2011; Morris et al., 2010).

Most of these personal and mobile healthcare applications have focused on what we in this chapter will label as "single-loop" applications. In single-loop applications, the treatment is focused mainly on patient self-management of well-being, healthy living, and/or disease care and treatment. The loop in this approach contains three main steps: (1) the phone, or another embedded system component, is used for manual or automatic monitoring of important parameters; (2) based on collected data, the system builds a model reflecting the current state of the user's health-related condition; and (3) based on this model, the system provides feedback to the user aimed at enhancing a general awareness and to promote and encourage behavioral changes, which in the long run foster a more healthy living (see [Bardram and Frost, 2016] for a review of the personal health technology design space).

As a good example of this single-loop approach, the BeWell system (Lane et al., 2011) targets end-user self-management of well-being with three distinct phases as illustrated in Fig. 10.1 (left). First, everyday behaviors are automatically monitored. Next, the impact of these lifestyle choices on overall personal health is quantified using a model of well-being. Finally, the computed well-being assessment drives feedback designed to inform the user about his or her health status and to promote healthy behavior. Specifically, the BeWell smartphone

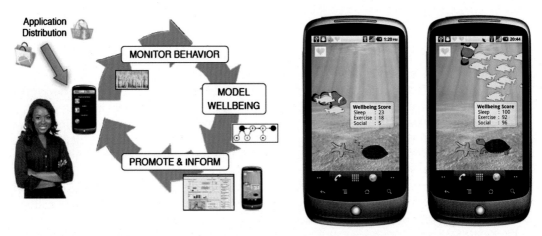

FIGURE 10.1 The BeWell system. Left: The BeWell loop supports end-user self-management of well-being with three distinct phases: (1) everyday behaviors are automatically monitored; (2) the impact of lifestyle choices on overall personal health is quantified using a model of well-being; and (3) the computed well-being assessment drives feedback designed to inform the user about his or her health status. Right: The BeWell smartphone user interface providing feedback using an aquarium as a metaphor.

application is monitoring and inferring three types of wellness behavior: sleep, physical activity, and social interaction. Motivational feedback to the user is designed as a graphical wallpaper on the smartphone using an aquarium as a metaphor as shown in Fig. 10.1 (right); the movement of a turtle, a clown fish, and a school of yellow fish is reflecting sleep, physical activity, and social interaction, respectively.

The main motivation for using this single-loop design is that such systems should empower the patient to be able to handle his or her own wellness and/or disease. Besides the benefit of empowering the patient, this model for healthcare delivery also contains huge practical and economical benefits at a time where the modern healthcare system is facing significant challenges due to an increase in age, number of chronic and lifestyle-related diseases (like obesity, diabetes, and COPD), and shortage of clinical staff members (nurses as well as doctors). Therefore, designing personal health technology applications that promote healthy living, help the patient to care for him or herself, and in general provide the patient with greater self-awareness about his or her health condition is a serious part of the answer to these challenges.

However, these pervasive healthcare systems are often used by patients who are part of a clinical treatment and it is crucial to design these systems to also incorporate the clinical part of the treatment. It is not feasible—or even realistically doable—to rely on patient self-monitoring and self-care; most patients with a moderately severe health condition will be associated with some clinical treatment. As such, personal health technology should be designed with clinical treatment in mind and then be used to increase the quality and efficiency of the clinical treatment of the patient. In this chapter we call this approach the "double-loop" design of personal healthcare applications; the first loop is between the patient and the system and the second loop is between the system and the clinic, as illustrated in Fig. 10.2.

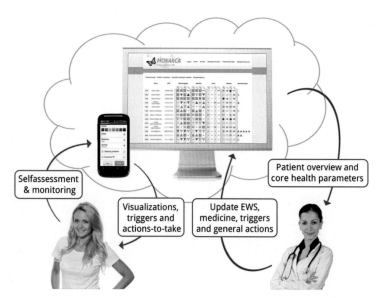

FIGURE 10.2 The double-loop setup in the MONARCA project. The first loop takes place between the patient and the smartphone health application. The second loop takes place between the clinician and the clinical portal. This double-loop setup mediates the relationship between the patient and his or her clinical treatment.

However, designing a double-loop, rather than a single-loop, personal health system is far more complex since this is not merely a user-experience and a technological design challenge but a socio-technical design challenge as well. A core tenet in the double-loop approach to personal health technology is the assumption and expectation that clinical care and treatment around a specific chronic condition is (re-)organized around the technology. This means that—ideally—technology and work organization is codesigned in order to leverage the opportunities that both the new technology provides as well as the opportunities that the organization provides, as well as accommodates the weaknesses and drawbacks of both technology and work organization.

3. CASE: DESIGNING FOR DOUBLE-LOOP TREATMENT IN MENTAL HEALTH

The overall goal of the MONARCA project is to make the treatment of patients suffering from affective disorder more efficient (Gravenhorst et al., 2015). This is done by supporting patients to self-report health data, which can be continuously reviewed by clinicians. The main usage scenario is shown in Fig. 10.3.

By using a personal smartphone-based healthcare application, the patient is provided with a greater awareness of his or her disease and can exercise a much greater degree of self-care and self-treatment. The system lets the patient self-assess and review a number of health parameters and supports illness management. For example, patients can use the data to determine adherence to medications, investigate illness patterns, and identify early warning signs (EWSs) for upcoming affective episodes or test potentially beneficial behavior changes. Data collected can be used to predict and prevent the relapse of critical episodes. Through monitoring and persuasive feedback, the system helps patients implement effective short-term responses to warning signs and preventive long-term habits. This reduces the need for clinical supervision, treatment, and care, while at the same time empowers the patient in personally dealing with the disease.

By having access to the long-term patient monitoring data in the clinical portal, clinicians like a psychiatrist, psychologist, or nurse have a much more detailed insight into the patterns of a patient's disease as it unfold over time and in daily life. This allows for a much more efficient and targeted treatment of each patient. Moreover, by continuously monitoring core health parameters from a pool of patients, the clinic can proactively spot trends and identify patients who may be in need for clinical treatment. For example, if a patient is showing an

FIGURE 10.3 The core usage scenario in the MONARCA system: (1) the patient fills in self-assessment data on the smartphone (mood, stress, activity, etc.); (2) the psychiatrist monitors his patients on the clinical portal; (3) the patient and psychiatrist use historical data in the system during psychotherapy.

increasing depressive trend, this patient could be invited for an outpatient consultation with a psychiatrist before his or her depression becomes severe and life-threatening. Overall, the approach in the double-loop treatment setup is to transform treatment in mental health from a *reactive* model in which patients are admitted to treatment when a crisis has occurred to a more *proactive* treatment model, which allows the clinic to spot upcoming episodes and crisis before they emerge.

3.1 Design Process

The design of the MONARCA system was done in a user-centered and participatory design process applying the Patient-Clinician-Designer Framework (Marcu et al., 2011). As shown in Fig. 10.4, a group of patients, psychiatrists, psychologists, and designers held biweekly design meetings for a period of 6 months. During these design meetings, both the interaction design of the smartphone app and the clinical portal was designed, as well as how the system should be implemented in clinical practice. As such, this was a user-centered design of the socio-technical system setup trying to consider both the technical as well as the organizational aspects.

As illustrated in Fig. 10.3 the socio-technical system design pivots around a core usage scenario in which the patient would collect and enter behavioral and disease relevant data on the smartphone, which then would be available on the clinical portal. The assumption in the scenario was that the psychiatrist responsible for the patient would "on a regular basis" consult the clinical portal and monitor how the patient was doing and take appropriate action, if needed. Moreover, a core design assumption was that the psychiatrist (or psychologist) would use historical data during outpatient consultations to discuss trends and patterns with the patient while also using "positive moments" as leverage in further treatment. In CBT, e.g., a common strategy is to help a depressed patient to think back on positive moments in life and reflect upon what helped during this period. With the MONARCA system, it would be straightforward to point to the historical mood chart and point to a positive mood period and ask what worked for the patient during this period.

FIGURE 10.4 The user-centered design workshop involving a group of clinicians, patients, and designers for biweekly design meetings.

3.2 Smartphone Application

The main goals of the MONARCA phone application are to (1) provide an input mechanisms for patients to fill in their self-assessment data; (2) collect objective sensor data from the phone; (3) provide a simple historic visualization of the data entered; (4) provide feedback and suggest actions to take in situations that presents risks; and (5) help patients to keep track of their prescribed medication. The main reason for using a mobile phone application is that the phone is always with the patient. This is useful not only for the objective activity sensing but also for collecting the self-assessment data, since the phone is much easier available than a web browser. Fig. 10.5 shows the user interface design of the main screens in the MONARCA application consisting of a home screen, linking to five different subscreens: (1) self-assessment, (2) visualizations, (3) triggers and actions to take, (4) medicine and (5) settings.

Significant effort was put into designing the self-assessment form on the phone as simple and concise as possible. A core requirement from the patients involved in the user-centered design process was that the list of self-reported items should be as short as possible, so that the self-assessment could be done quickly. Based on this input the MONARCA self-assessment

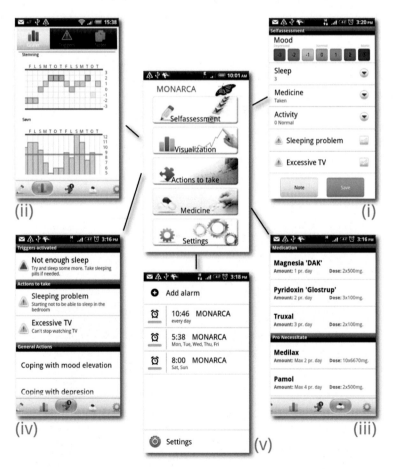

FIGURE 10.5   The MONARCA Android application user interface.

form as shown in Fig. 10.5 (i) contains a minimum set of things to monitor, which are divided into a set of *mandatory* self-assessment items, which is absolutely crucial to collect over time in the treatment of a bipolar patient and a set of *optional* self-assessment items, which can be additional data that are relevant for a specific patient. For bipolar disorder patients, mandatory self-assessment items include mood, sleep, activity, and medicine adherence. Optional self-assessment items can be configured for each patient and include EWSs, alcohol, and stress. A free-text note can always be added to the self-assessment. On a daily basis, an alarm on the phone reminds the patient to report self-assessment data, which can be entered and edited up until midnight. In addition to the self-assessment data, the phone automatically samples objective data. This includes *physical activity* data as measured by the accelerometer or pedometer in the phone, *social activity* as measured by the number of incoming and outgoing phone calls and text messages, *mobility* as measured by the phone's location service (e.g., GPS), and general *phone usage* as collected from the phone's system log.

Once data are reported, a historical overview is visualized directly on the phone, as shown in Fig. 10.5 (ii). This visualization is intended to provide patients with an awareness of the historical development of their disease. The system also provides more active feedback to the patient. As illustrated in Fig. 10.5 (iv) the system has three types of recommendation:

- *Triggers*—simple if-then rules giving an advice in a certain circumstance. For example, if a patient is sleeping less than 5 h in 4 successive days, the system can suggest him to sleep more regularly and e.g., take a sleeping pill.
- *Actions to take*—simple advice of how to handle a EWS. For example, if a patient starts sleeping in the living room rather than the bedroom, this can be a EWS of a manic period. In this case the patient should often consult his doctor and start taking special medication for manic symptoms.
- *General actions*—these are more general advice targeted to the patient. For example, a young patient might want to call his mother in case a depression is coming up.

Finally, the application shows the patient's list of prescribed medicine, as shown in Fig. 10.5 (iii). Medication is managed and updated by the clinic via the clinical portal.

3.3 Clinical Portal

The clinical portal (a website) is used by clinicians to review patient health data as entered and collected on the smartphone. Clinicians can manage personal triggers, EWSs, medication, and in general configure the system. When logging in, clinicians get a dashboard providing an overview of their patients (name and social security number) and how they are doing on the core parameters of self-reported mood, activity, sleep, and medicine adherence. These health parameters are shown for the last 4 days in order to visualize any trends in the data, for the clinicians to spot the ones in need of urgent attention. Moreover, any activated triggers or EWSs are shown on the right, visualized by orange and red warning triangles. This dashboard is shown in Fig. 10.6 (top).

In the dashboard, the clinician can select a single patient and review detailed health and behavioral data on a separate page as shown in Fig. 10.6 (bottom). This page contains all the data sampled from the smartphone as graphs. In general, the clinician has access to reviewing and managing data and configurations for each patient, including self-assessed health data; automatically sampled data; medicine prescriptions; EWSs; actions to take; general actions;

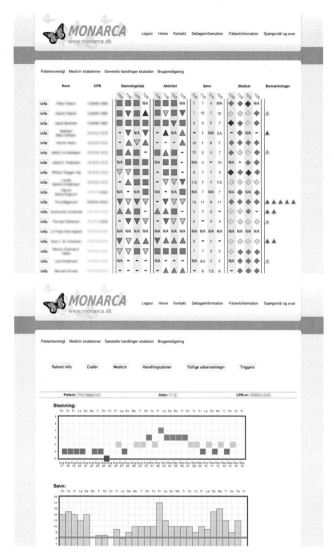

FIGURE 10.6 The Clinical portal. Top: the "Dashboard" showing a list of patients for a specific doctor or nurse. Each line is a patient (name and ID number in the left columns), showing mood, activity, sleep, and medicine data for the last 4 days. Activated triggers and EWSs are shown in the far right column. Bottom: the "Patient Data" page used by clinicians to review health data for a specific patient. This picture shows mood and sleep data.

and triggers. In addition, a small medical diary containing a simple medical record for each patient can be entered into the clinical portal.

3.4 Clinical Implementation

The MONARCA system was designed and deployed in the clinic according to the double-loop model illustrated in Fig. 10.2. Continuous mood tracking and graphing, controlling EWSs and triggers, activity logging, and medication compliance training are core ingredients in CBT for the experienced, but not yet stable for bipolar disorder patient (Basco and Rush, 2005). Hence, the treatment loop forms a basis for continuous self-treatment and self-care by experienced patients suffering from bipolar disorder. The system was therefore handed out to patients in the clinic who had been fully diagnosed and had been undertaking some degree of psychoeducation prior to the use of the system on his or her own. The system was hence given to patients being discharged from the clinic or as part of an ongoing outpatient treatment.

As explained earlier—and illustrated in Fig. 10.3—the system was designed to allow psychiatrists to monitor their own set of patients on a daily basis and then contact patients, as needed. For example, if a patient was showing persistent depressive traits over a longer period, or constantly triggering EWSs, this patient should be contacted. Moreover, the clinical portal with its patient data overview (Fig. 10.6) was designed for clinicians like psychiatrists and psychologist to be used during outpatient consultations. However, during the deployment of the MONARCA system, the work practice at the clinic evolved into a "call center" setup in which a trained nurse was monitoring all patients associated with the clinic, instead of each psychiatrist overlooking his or her own patients. The number of patients enrolled in the MONARCA system treatment is never more than 40, and the nurse is able to monitor and keep an overview of them all. The nurse monitors the patients on a daily basis, and based on instructions from the head psychiatrist and her own training, she is able to judge which of the patients to contact. Contact happens either through text messaging or by calling the patient. If the nurse concludes that a patient case should be reviewed by a psychiatrist, she notifies the patient's psychiatrist through email, attaching a printout of the data from the MONARCA system, and asks the doctor to make a decision as to whether the patient should come to the clinic. In this case, a visit is scheduled, typically by a secretary.

4. LESSONS LEARNED

The MONARCA system has been subject to both usability and clinical studies. Usability studies during clinical trials have shown that patients found the system very useful and easy to use, and we found a significant uptake in the use of the system; the adherence rate of the system was in average 87% as measured in the number of days patients entered their self-assessment (Bardram et al., 2013). Clinical studies show that there is a significant correlation between self-rated mood and clinically rated depression and mania (Faurholt-Jepsen et al., 2015b). This result indicates that patients are able to assess their own disease and that self-rated mood is hence a valid indicator for treatment in the clinic. Moreover, studies show that there is a correlation between "social activity" (as measured in terms of number of incoming/outgoing calls and text messages on the phone) and depression and mania (Faurholt-Jepsen et al., 2016). However, a randomized clinical trial (RCT) showed no significant effects of daily self-monitoring using the MONARCA system on depressive or manic symptoms (Faurholt-Jepsen et al., 2015a).

In total, the MONARCA system has been deployed for more than 4 years and has been used extensively in clinical trials involving more than 150 patients. During this period of deployment, a number of lessons of a more socio-technical nature have emerged, which is the topic of this chapter. These lessons can roughly be divided into four areas related to (1) using the system as part of clinical treatment; (2) reorganizing clinical treatment to include the use of the system; (3) socio-technical scalability of the system; and (4) socio-technical integration. In the following, we shall present and discuss these four findings.

4.1 Using Personal Health Technology

As outlined earlier, the MONARCA system was carefully designed in a participatory user-centered design process involving designers, patients, and clinicians, including psychiatrists, psychologists, and nurses. As illustrated in Fig. 10.3, the system was designed around a core usage scenario in which the patient would collect and enter behavioral and disease relevant data on the smartphone, which was continuously monitored by the responsible psychiatrist and used actively during consultations and CBT sessions. This process is depicted more formally in the swimlane model in the top part of Fig. 10.7: A patient installs the app on his or

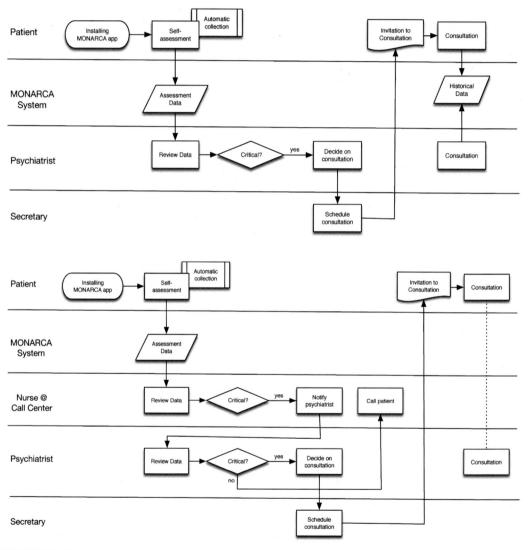

FIGURE 10.7 Swimlane models showing processes mapped to actors. Top: the original scenario shown in Fig. 10.3. Bottom: the revised scenario, in which a nurse in a call center is handling the monitoring of the patient.

her phone and engages in self-assessment of data while the app automatically collects data in the background. All data are synchronized with the MONARCA system. The psychiatrist is reviewing data on a regular basis and if he finds critical data on a patient, she/he might decide to call the patient for an outpatient consultation. This is handed over to a secretary, who is responsible for scheduling an appointment. During the consultation, the patient and the psychiatrist review the data together.

However, during the deployment of the system, it was decided that rather than having psychiatrists or psychologists to use the system, it should be deployed in the "call center" setup in which a nurse would be the main point of contact. There were several reasons for this change to the original process. First, clinicians do not use computers during consultations with patients. The kind of computers available in the clinic are old desktop computers, which are slow and not very useful, and are rarely available in consultation rooms. Therefore there were a lot of practical hardware problems of getting access to the MONARCA system during consultations. Moreover, the MONARCA system was not integrated with the other computer systems at the clinic—such as the electronic medical record (EMR) or the medicine system—and would hence require a clinician to maintain medical information in several systems. Second, the idea that clinicians should be monitoring patients turned out to be overly naïve. Clinicians see a large number of patients and there is no way that a clinician can monitor all these patients on a daily basis. Moreover, due to the type of hardware available and its accessibility, there was no computer on which a clinician could do this monitoring. Only few senior clinicians—such as senior consultants and psychiatrists—had an office with a PC. Most other clinicians did not have access to a PC. As such, the monitoring part of the original scenario was simply not feasible to implement in practice. Similarly, since there was computer available in consultations rooms, the MONARCA Web portal could not be accessed during outpatient consultations. Third, the responsibility of the treatment of an outpatient using the MONARCA system turned out to be unclear. Clinicians at the clinic are responsible for a patient during hospitalization and outpatient consultations. They are, however, not in the same manner responsible for a patient in-between consultations; this is the responsibility of the patient's general practitioner. As such, the notion that a clinician would "continuously monitor" the disease development of a patient turned out not to be applicable, since it would extend far beyond the current work responsibility of the clinicians.

Due to these constraints—which became apparent as part of the deployment and after the system was designed and developed—another setup was proposed and implemented. In this setup, a nurse would be full-time allocated to oversee patients in the MONARCA system and hence be responsible for monitoring of and communication with all patients. The nurse would then—if needed—contact a clinician. This process is shown in the lower part of Fig. 10.7: The patient installs and uses the MONARCA app as before, but now a nurse is reviewing the data on a continuous basis. If the nurse finds that data on patient is critical, she/he notifies the psychiatrist and hands over a printout from the MONARCA system. If the psychiatrist finds the data to be critical as well, she/he will—as before—ask the secretary to schedule a consultation with the patient. If not, the psychiatrist may ask the nurse to call the patient in order to double check. During the outpatient consultation, the psychiatrist and patient do not have access to the MONARCA system via the Web portal.

In effect, a small call center was established. No psychiatrists or psychologists used the system and no one asked for access to it, even though they could. This had the very unfortunate consequence that most patients were frustrated that their clinician had not looked into the data that they meticulously had entered into the system, when meeting for consultations.

In retrospect, the deployment of the MONARCA system revealed a set of core socio-technical challenges and issues, including systems and technical ones related to the lack of useful hardware in the intended usage situations (old PCs only available in senior clinicians offices and no useful hardware available during patient consultations) and lack of system integration to the EMR back-end systems, as well as organizational issues related to the distribution of responsibility for patients over time and in-between visits to the clinic.

4.2 Reorganizing Socio-technical Systems

The first version of the MONARCA system was deployed during 2011, and the system has been in use for consecutive 5 years, while being part of four clinical studies and pilot tests. The system has, however, not been put into wider clinical use for all mood disorder patients in the entire HMO. Why not? This is because putting into operation a system like MONARCA is a rather significant socio-technical reorganization of the treatment which comes with big costs both technically and organizationally. Technical costs are associated with the licensing and operating the system and the organizational costs are associated with implementing a reorganization of the work in the different clinics in the HMO. Therefore, in order to make a decision of taking into use such a system, decision-makers are requesting evidence for the benefit of the system. Following the HIMSS STEPS model,[3] such benefits could include increased satisfaction (S), improved treatment (T), increased electronic data quality (E), increasing prevention (P), and/or economical savings (S). Establishing evidence for these parameters in a sound manner is by no means a simple task. For example, in order to establish clinical evidence for the efficacy of treatment of bipolar patients using the MONARCA system setup, a RCT has been conducted (Faurholt-Jepsen et al., 2013). Conducting such an RCT is a significant effort and investment. In the MONARCA RCT a total of 78 patients were included over a 30-month period, each using the system for 6 months and each treated by their psychiatrist. In addition, a psychiatrist was hired full-time to conduct and manage the RCT over the entire period. As such, huge efforts of many people have gone into this RCT in order to just establish clinical evidence for improved quality in treatment—and this RCT did not touch any of the other STEPS parameters.

A core lesson learned from the MONARCA project is that socio-technical reorganization of clinical pathways to incorporate personal health technology is a very slow and long-term process. This is due to the fact that decision-makers in healthcare seek to rely on evidence-based decisions; clinical pathways and guidelines are designed based on clinical evidence in international peer-reviewed literature. And since a typical RCT takes 4–5 years in terms of planning, funding, execution, analysis, and publications, evidence for new technology is established only over a very long period of time. Overall, the following phases are needed in order to bring a new personal health technology from idea and design to implementation in clinical practice: we are looking at time frames like:

- 2–3 years—design and development of personal health technology
- 3–4 years—establishing evidence via an RCT design, execution, and publication
- 1–2 years—socio-technical implementation of the technology in operation and organizational redesign of clinical guidelines and pathways.

[3] See http://www.himss.org/.

Hence, in total we are looking at a time frame of 6–9 years from initial design of a system to implementation in clinical practice—at best. And, since technological development is progressing at a more rapid pace, the system will be completely outdated—at least from a technological point of view—by the time it is implemented in clinical and organizational practice.

4.3 Socio-technical Scalability of Personal Health Technology

Similar to other personal health technologies (cf., Bardram and Frost, 2016), the MONARCA system is designed for large-scale use by potentially millions of patients. As such, the technical system architecture consisting of powerful smartphones as a front-end sensor and application platform and high-performance cloud computing for back-end data collection, management, processing, and visualization provide a highly scalable technical setup. However, such a system is no longer considered solely as a technical system, but as a socio-technical system, a core question becomes what the impact the social dimension has on the scalability of the system? Based on the experience from the MONARCA project, we observed scalability issues both in the patient's part of the system (the patient loop) as well as in the clinical part of the system (the clinical loop).

From the patient's point of view, the main scalability concern was related to the link back to the clinic. The usability studies of the system have consistently shown that a crucial aspect of the usefulness of the system was tied to the fact that the patient knew—or presupposed—that a professional clinician was "listening" in the other end. Based on the statement of a patient, this connection back to the clinic was viewed as a "life jacket" to be used in case rescue was needed. However, it was quite unclear to the patient how this setup was scaling. There was no way to see if a clinician had seen the data, and it was unclear as to when and how often a patient's data would be monitored. For example, there was no indication in the user interface revealing if data or messages had be seen or when data were monitored (e.g., only during day shift or 24/7). Another scalability issue for patients was the communication links with the clinic. Originally, the system was designed to allow a patient to call or send a message directly to the clinic. But this was disabled in the system during deployment; there was a fear that this would just open up for a lot of communication. Hence, clinic–patient communication did not scale at all from an organizational/resource point of view. This points to a core contradiction to the Danish healthcare system and the use of personal health technology; essentially, the Danish healthcare system (like most other contemporary healthcare organizations) is designed to prioritize which patients get in contact with the healthcare system and especially the specialized doctors inside it. A fundamental design principle of a healthcare system is to protect clinicians (in order to allow them to concentrate of the few really ill patients) and to prevent—or prioritize/triage—other patients to stay outside. This is the reason for the complex referral process in place in all healthcare systems. Now, the MONARCA system completely short-circuits this setup, which has been crafted and refined over decades. Of course, if patients had direct access to clinicians—especially psychiatrists and psychologist—this would be ideal. However, this does not in any way scale.

In addition to these patient-related scalability issues, a set of related clinical scalability issues arose. A core question that came up early was how many patients can the call center setup handle? How many patients can a nurse handle? And should there be a traditional front-line/back-line setup in which a group of front-line nurses take all calls and redirect severe case to a back-line senior nurse or psychiatrist? A related question was related to the scalability of the monitoring of patients; how detailed should patients be monitored and how often? The number of patients that one nurse could monitor was highly dependent on the level of details (e.g., number of data points) and the frequency (e.g., once a week, once a day, or several times a day). Related to this, was the question as to how the nurse should act; what was the threshold for acting and exactly what should she/he do? For example, are moderate depression scores for 3 days in a row something that the nurse should act upon? And if so, how? By sending a message, by calling, or by contacting a psychiatrist? Moreover, a basic issue for the nurse was the recurring question of how well the data in the system could be trusted. Patients are definitely not alike, and they fill in the self-assessment scores in very different ways; a "$\frac{1}{2}$" in mood score for one patient may be a "2" for another. Hence, the nurse reported that it was very important to get to know each of the patients individually in order to understand the data and when to react. However, such a personalized interpretation of data for each patient does not scale particularly well. Similar questions were asked regarding the automatically collected data, reflecting physical activity, social activity, mobility, and phone usage. For example, it is well-known that automatic sensing of physical activity is error-prone (e.g., in a Danish context bicycling is often not captured) and this automatically collected data were again subject to personalized interpretation of the nurse, if used at all.

Observations like the ones outlined here emphasize that scalability of personal health technology is not a technical issue alone but is of an inherent socio-technical nature; in fact, social scalability seems harder to be achieved than technical scalability. And the design of the human–computer interface seems to be sitting at the borderline between the social and technical design of the system; just because the system scales well technically, it might not scale well usability wise. For example, in the MONARCA project once that the call center deployment model was decided, careful design of the clinical dashboard became essential for socio-technical scalability. The design of this dashboard—with all the details of what order to list patients, what data to show, with which colors and icons, etc.—became absolutely crucial in determining how many patients a nurse could handle.

4.4 Socio-technical Integration of Personal Health Technology

At some point, some psychiatrists and psychologists became interested in access to the system and would like to use it during patient consultations—as originally designed for. However, since desktop computers are unfit for patient–clinician consultation, it was decided to use iPads instead. From the MONARCA system's point of view, the clinical portal would run perfectly fine on an iPad (it is a responsive Web site) so this seemed like a path forward (or back) to the originally intended deployment setup. However, due to technical constraints and security requirements on the wireless network in the hospital, each iPad had to be registered and allowed access to the WiFi. And figuring out how this approval process worked, who was responsible, and what to do was not a straightforward issue. In essence, the MONARCA project was fighting for some very limited resources in terms of network administrators in the IT department, which were completely overbooked by a dysfunctional WiFi network.

Moreover, deploying a personal health technology as a part of a HMO might require the system to be hosted and operated by the HMO. This was the case of MONARCA; the IT department of the HMO required the system to be hosted by them for a number of reasons: security, privacy, data protection, ownership and responsibility, legal, and political (the IT department had a policy of hosting everything themselves). This hosting setup, however, goes completely against the assumptions in most personal health technologies, which are designed according to a cloud-based architecture in which the physical location of hosting becomes irrelevant since this just happens "in the cloud". This hosting setup also implied a lot of nonfunctional requirements in order for the system to fit into the infrastructure of the HMO: special requirements for operations, logging, warning system, selection of operating system (OS), and database technology, etc. For example, the MONARCA system was implemented to run on Linux for high-availability and scalability using a CouchDB for data synchronization. However, the IT department of the HMO would only host technology on a Microsoft platform using Windows Server OS and MS SQL Server.

The point of these two examples are that the socio-technical design of personal health technology may need to take into consideration the integration into the technical infrastructure of e.g., a hospital or an HMO, and that this infrastructure again is a socio-technical infrastructure with a number of intermixed organizational, regulatory, legal, political, and technical requirements and constraints.

5. CONCLUSION

Personal health technology is rapidly emerging as a response to the challenges associated with significant increase in chronic noncommunicable diseases. The overall design paradigm behind most of these applications is to manually and automatically sample data from sensors and smartphones and use this to provide patients with an awareness of their illness and give recommendation for treatment, care, and healthy living. Few of these systems are, however, designed to be a part of a complex socio-technical care and treatment processes in existing healthcare systems and clinical pathways. In this chapter, we presented a case of designing personal health technology for mental health, which is integrated into hospital-based treatment. This system helps patients to manage their disease by tracking and correlation behavior and disease progression and provide feedback to them, while also being deployed as part of a clinical outpatient treatment. Hence, clinicians are "in the loop" and can monitor and provide feedback to patients. This socio-technical setup was named the "double-loop" treatment setup and is illustrated in Fig. 10.2. The chapter presented a set of four lessons learned from running several clinical trials spanning a 4-year deployment period. These lessons were related to (1) using the system as part of clinical treatment; (2) reorganizing clinical treatment to include the use of the system; (3) socio-technical scalability of the system; and (4) socio-technical integration.

The first lesson was that a user-centered, participatory design process is not sufficient in the design of socio-technical systems. Even though the socio-technical usage setup of the system was designed as part of the design process—i.e., that clinicians should use the system for continuous monitoring of patients and during consultations—this setup was never adopted and implemented during the actual deployment for all sorts of "practical" reasons, including lack of proper hardware, network access, time, and skills. Instead, the system was deployed in another socio-technical configuration—that of a "call center"—which imposed a set of new design requirement on the system, including the design of the clinical dashboard. As such, real-world constrains can be hard to predict during design, and the socio-technical configuration of these personal health technologies should hence be able to adapt accordingly.

Second, adoption of personal health technology in a managed (public and private) healthcare organization has turned out to be a very long process in order to establish the necessary clinical evidence for the efficacy of the technology combined with necessary organizational implementation and reorganization needed, in order to take advantage of the technology. In the MONARCA case, the initial design started in 2010, and it is not yet fully deployed in production. Hence, a 7–10 year time frame from initial design to complete socio-technical uptake of these technologies may be expected.

Third, despite the fact that the MONARCA system scales well from a technical perspective, we found that scalability was an issue as to the social/organizational part of the system, which applied both for the patient loop as well as for the clinical loop. From the patient's perspective the system was considered a "life jacket" and patients trusted that clinicians would monitor their mental health directly, while there was a limitation as to how many patients the nurses in the call center could actually handle. We hence found that personal health technology scales poorly from a socio-technical point of view if it short-circuits the healthcare system and opens up a direct line of communication between patients and clinicians. Hence, the socio-technical design of personal health technologies needs to support more traditional instruments for handling patients at scale in a healthcare system, including referral, prioritization, and triage mechanisms.

Fourth, the implementation and use of a personal health technology in a HMO requires a socio-technical integration of the system into the technical infrastructure. Hence, rather than being designed as a stand-alone system used for patients alone, the system needs to be designed to be integrable according to the constrains of the specific HMO's technical infrastructure, which again is a socio-technical infrastructure with a number of intermixed organizational, regulatory, legal, political, and technical requirements and constraints.

Personal health technology holds great promise to address some of the current and future challenges in chronic disease management and treatment. But it is essential to its success that implementation adhere to an efficient socio-technical setup. As a reminder of this, we have proposed the MONARCA double-loop treatment setup and discussed how the design of socio-technical clinical use, reorganization, scalability, and integration has to be taken into consideration when implementing personal health technology as part of clinical treatment and care.

Acknowledgments

The research presented in this chapter has been partially supported by the FP7 European Framework program as part of the MONARCA project[4] and by the Innovation Fund Denmark as part of the RADMIS project.[5]

References

Arnhold, M., Quade, M., Kirch, W., 2014. Mobile applications for diabetics: a systematic review and expert-based usability evaluation considering the special requirements of diabetes patients age 50 years or older. Journal of Medical Internet Research 16 (4), e104.

Arnrich, B., Mayora, O., Bardram, J., Troster, G., 2010. Pervasive healthcare – paving the way for a pervasive, user-centered and preventive healthcare model. Methods of Information in Medicine 1, 67–73.

[4] http://monarca-project.eu/.

[5] http://www.cachet.dk/research/projects/RADMIS.

Bardram, J.E., 2008. Pervasive healthcare as a scientific discipline. Methods of Information in Medicine 3 (47), 129–142.

Bardram, J.E., Frost, M., 2016. The personal health technology design space. IEEE Pervasive Computing 15 (2), 70–78.

Bardram, J.E., Frost, M., Szántó, K., Faurholt-Jepsen, M., Vinberg, M., Kessing, L.V., 2013. Designing mobile health technology for bipolar disorder: a field trial of the MONARCA system. In: Proceedings of the SIGCHI Conference on Human Factors in Computing Systems. ACM, New York, NY, USA, pp. 2627–2636.

Bardram, J.E., Frost, M., Szanto, K., Marcu, G., 2012. The MONARCA self-assessment system: a persuasive personal monitoring system for bipolar patients. In: Proceedings of the 2nd ACM SIGHIT International Health Informatics Symposium. ACM, New York, NY, USA, pp. 21–30.

Basco, M.R., Rush, A.J., 2005. Cognitive-behavioral Therapy for Bipolar Disorder, second ed. The Guilford Press.

Bauer, J., Consolvo, S., Greenstein, B., Schooler, J., Wu, E., Watson, N.F., Kientz, J., 2012. ShutEye: encouraging awareness of healthy sleep recommendations with a mobile, peripheral display. In: Proceedings of the SIGCHI Conference on Human Factors in Computing Systems. ACM, New York, NY, USA, pp. 1401–1410.

Bauer, M., Grof, P., Rasgon, N., Glenn, T., Alda, M., Priebe, S., Ricken, R., Whybrow, P.C., 2006. Mood charting and technology: new approach to monitoring patients with mood disorders. Current Psychiatry Reviews 2 (4), 423–429.

Burns, M., Begale, M., Duffecy, J., Gergle, D., Karr, C., Giangrande, E., Mohr, D., 2011. Harnessing context sensing to develop a mobile intervention for depression. Journal of Medical Internet Research 13 (3).

Cafazzo, A.J., Casselman, M., Hamming, N., Katzman, K.D., Palmert, R.M., 2012. Design of an mHealth app for the self-management of adolescent type 1 diabetes: a pilot study. Journal of Medical Internet Research 14 (3), e70.

Consolvo, S., Klasnja, P., McDonald, D.W., Landay, J.A., 2014. Designing for healthy lifestyles: design considerations for mobile technologies to encourage consumer health and wellness. Foundations and Trends in Human-Computer Interaction 6 (3–4), 167–315.

Consolvo, S., McDonald, D.W., Toscos, T., Chen, M.Y., Froehlich, J., Harrison, B., Klasnja, P., LaMarca, A., LeGrand, L., Libby, R., Smith, I., Landay, J.A., 2008. Activity sensing in the wild: a field trial of UbiFit garden. In: Proceedings of the SIGCHI Conference on Human Factors in Computing Systems. ACM, New York, NY, USA, pp. 1797–1806.

Faurholt-Jepsen, M., Frost, M., Ritz, C., Christensen, E., Jacoby, A., Mikkelsen, R., Knorr, U., Bardram, J., Vinberg, M., Kessing, L., 2015a. Daily electronic self-monitoring in bipolar disorder using smartphones–the MONARCA I trial: a randomized, placebo-controlled, single-blind, parallel group trial. Psychological Medicine 1–14.

Faurholt-Jepsen, M., Vinberg, M., Frost, M., Christensen, E.M., Bardram, J.E., Kessing, L.V., 2015b. Smartphone data as an electronic biomarker of illness activity in bipolar disorder. Bipolar Disorders 17 (7), 715–728.

Faurholt-Jepsen, M., Vinberg, M., Christensen, E.M., Frost, M., Bardram, J., Kessing, L.V., 2013. Daily electronic self-monitoring of subjective and objective symptoms in bipolar disorder – the MONARCA trial protocol (MONitoring, treAtment and pRediCtion of bipolAr disorder episodes): a randomised controlled single-blind trial. BMJ Open 3 (7).

Faurholt-Jepsen, M., Vinberg, M., Frost, M., Debel, S., Margrethe Christensen, E., Bardram, J.E., Kessing, L.V., 2016. Behavioral activities collected through smartphones and the association with illness activity in bipolar disorder. International Journal of Methods in Psychiatric Research 25, 309–323.

Ferreira, P., Sanches, P., Höök, K., Jaensson, T., 2008. License to chill!: how to empower users to cope with stress. In: In Proc. NordiCHI 2008. ACM, New York, NY, USA, pp. 123–132.

Frost, M., Doryab, A., Faurholt-Jepsen, M., Kessing, L.V., Bardram, J.E., 2013. Supporting disease insight through data analysis: refinements of the MONARCA self-assessment system. In: Proceedings of the ACM International Joint Conference on Pervasive and Ubiquitous Computing. ACM, New York, NY, USA, pp. 133–142.

Gerteis, J., Izrael, D., Deitz, D., LeRoy, L., Ricciardi, R., Miller, T., Basu, J., 2014. Multiple Chronic Conditions Chartbook. Agency for Healthcare Research and Quality (AHRQ) Publications, Rockville, MD.

Gravenhorst, F., Muaremi, A., Bardram, J.E., Grünerbl, A., Mayora, O., Wurzer, G., Frost, M., Osmani, V., Arnrich, B., Lukowicz, P., Tröster, G., 2015. Mobile phones as medical devices in mental disorder treatment: an overview. Personal and Ubiquitous Computing 19 (2), 335–353.

Hoffman, C., Rice, D., Sung, H.-Y., 1996. Persons with chronic conditions: their prevalence and costs. JAMA 276 (18), 1473–1479.

Lane, N.D., Choudhury, T., Campbell, A., Mohammod, M., Lin, M., Yang, X., Doryab, A., Lu, H., Ali, S., Berke, E., 2011. BeWell: a smartphone application to monitor, model and promote wellbeing. In: Proceedings of the 5th International ICST Conference on Pervasive Computing Technologies for Healthcare (Pervasive Health 2011). IEEE Press.

Lee, H.R., Panont, W.R., Plattenburg, B., de la Croix, J.-P., Patharachalam, D., Abowd, G., 2010. Asthmon: empowering asthmatic children's self-management with a virtual pet. In: Proceedings of the 28th of the International Conference Extended Abstracts on Human Factors in Computing Systems. ACM, New York, NY, USA, pp. 3583–3588.

Lin, J., Mamykina, L., Lindtner, S., Delajoux, G., Strub, H., 2006. Fish 'n' steps: encouraging physical activity with an interactive computer game. In: Dourish, P., Friday, A. (Eds.), Proceedings of the ACM International Conference on Ubiquitous Computing, vol. 4206 of Lecture Notes in Computer Science. Springer, Berlin/Heidelberg, pp. 261–278.

Mamykina, L., Mynatt, E.D., Kaufman, D.R., 2006. Investigating health management practices of individuals with diabetes. In: Proceedings of the SIGCHI Conference on Human Factors in Computing Systems. ACM, New York, NY, USA, pp. 927–936.

Marcu, G., Bardram, J.E., Gabrielli, S., 2011. A framework for overcoming challenges in designing persuasive monitoring systems for mental illness. In: Proceedings of Pervasive Health. IEEE Press, pp. 1–10.

Matthews, M., Doherty, G., 2011. In the mood: engaging teenagers in psychotherapy using mobile phones. In: Proceedings of the SIGCHI Conference on Human Factors in Computing Systems. ACM, New York, NY, USA, pp. 2947–2956.

Morris, M., Kathawala, Q., Leen, T., Gorenstein, E., Guilak, F., Labhard, M., Deleeuw, W., 2010. Mobile therapy: case study evaluations of a cell phone application for emotional self-awareness. Journal of Internet Medical Research 12 (2), e10–12.

Nakonezny, P., Hughes, C., Mayes, T., Sternweis-Yang, K., Kennard, B., Byerly, M., Emslie, G., 2010. A comparison of various methods of measuring antidepressant medication adherence among children and adolescents with major depressive disorder in a 12-week open trial of fluoxetine. Journal of Child and Adolescent Psychopharmacology 20 (5), 431–439.

Pollak, J., Gay, G., Byrne, S., Wagner, E., Retelny, D., Humphreys, L., 2010. It's time to eat! Using mobile games to promote healthy eating. Pervasive Computing, IEEE 9 (3), 21–27.

Robertson, L., Smith, M., Tannenbaum, D., 2005. Case management and adherence to an online disease management system. Journal of Telemedicine and Telecare 11 (Suppl. 2), 73–75.

Siek, K.A., Connelly, K.H., Rogers, Y., Rohwer, P., Lambert, D., Welch, J.L., 2006. When do we eat? An evaluation of food items input into an electronic food monitoring application. In: Pervasive Health Conference and Workshops, pp. 1–10.

Smith, B.K., Frost, J., Albayrak, M., Sudhakar, R., 2007. Integrating glucometers and digital photography as experience capture tools to enhance patient understanding and communication of diabetes self-management practices. Personal and Ubiquitous Computing 11, 273–286.

World Health Organization, 2014. Global Status Report on Noncommunicable Diseases.

Designing Health Care That Works—Socio-technical Conclusions

Thomas Herrmann[1], Mark S. Ackerman[2], Sean P. Goggins[3], Christian Stary[4], Michael Prilla[5]

[1]Ruhr University of Bochum, Bochum, Germany; [2]University of Michigan, Ann Arbor, MI, United States; [3]University of Missouri, Columbia, MO, United States; [4]University of Linz, Linz, Austria; [5]Clausthal University of Technology, Clausthal-Zellerfeld, Germany

The socio-technical perspective on health care investigates the analysis, design, implementation, and adaptation of systems that incorporate both the technical and the social. The socio-technical perspective necessarily includes both technical functionality and social interactions between people in their various roles and activities.

The chapters of this book reflect established practices and emergent issues in socio-technical design for health care management and engineering. This chapter aims to synthesize key insights to characterize aspects of a socio-technical approach. The aim of this volume is to guide a more deliberate design of socio-technical health care systems.

This conclusion is based not only on the book's chapters but also draws from a workshop in August 2016 with the authors and editors.

1. THE EXTENDED VIEW OF A SOCIO-TECHNICAL PERSPECTIVE

Many chapters expound on the extended view that a socio-technical perspective affords. Below we discuss some of the most important issues when framing an analysis from a socio-technical perspective.

1.1 Increased Scope of Social Interaction

The socio-technical systems of health care involve increasingly complex social arrangements, and in turn, the social arrangements for patients' health care and clinicians' work more and more are parts of increasingly complex socio-technical systems.

Moreover, the types and roles of interacting people to be taken into consideration are not restricted to health care professionals and patients. Health care networks now can include home-helpers, family caregivers (Abru Amsha and Lewkowicz, 2017), friends, and people with the same health-related interests or problems being reached via the Internet. Other roles include mediators between physicians and the patients such as the cancer navigators who offer individual support to patients throughout cancer treatment (Jacobs and Mynatt, 2017). Yet other connections may help perform translation work between medical professionals and patients (Kazunias et al., 2013).

Furthermore, if not only phases of medical diagnosis and treatment but all kinds of activities that contribute to a healthy life are considered, relationships within and among families come into consideration. Parker et al. (2017) examines, for example, creating healthy behaviors within neighborhoods and communities. Consequently, the interplay between formal roles and informal roles in health care is being re-organized. New care actors have to be integrated and there will be an ongoing negotiation about the duties and rights of those informal roles (Jahnke et al., 2005).

A socio-technical perspective takes into account how the social environment is constantly changing as well. For example, with respect to social interaction it has to be understood that patients and the people in their environment are taking more and more responsibility in managing their own wellbeing and recovery. We might expect new responsibilities to emerge. For example, if patients are continuously monitored or if they convey data about themselves, either the patients may have to take responsibility or new professional roles may arise. In these cases, the responsibilities of handling this data will need to be clarified (Bardram and Frost, 2017).

Socio-technical considerations in health care, therefore, must include networks of people, technologies and systems, information and data brokering, changing roles, the consideration of physical space, the influence of family and the role of local communities, as well as disease- or condition-centered online communities among many other considerations.

Indeed, as people's lives and health become entangled in complex socio-technical networks, how we even view health care is evolving. As patients, individuals identify according to some disease or condition. However, today, health care is more than medical diagnosis and treatment. Health care includes all kinds of activities that contribute to a healthy life (Parker et al., 2017), and socio-technical systems may be designed to actively motivate individuals to live healthier lifestyles.

Health care work is exceedingly complex, and designing systems to incorporate it or improve it is just as complex. Many of the chapters in this book have emphasized the perspectives of practice and technology design. Practice theory (Wulf et al., 2011; Cetina et al., 2005) focuses on how complex work and interactions are entangled in the specifics of situations, including the specifics of technologies. Our version of the socio-technical viewpoint includes the messiness in and ambiguity of designing technical systems (Ackerman, 2000) to aid health care work, for patients or caregivers or clinicians.

To this point, one might read the synthesis of the chapters in this book as a systematic documentation of massive change, snapshotted at a point in time and rather difficult to direct. That specific concern, however, is why our focus is ultimately on the *design* of socio-technical systems for health care. Whether you fret about an over-emphasis on technological determinism (Kling, 1996) or embrace the inevitability that technology can significantly determine outcomes in socio-technical systems (Schroeder, 2007), you likely recognize, at least, that we need some way to link the design of technology and the changes in organizational structures and culture, human behavior, and care that are envisioned and desired. When the chapters in this book talk of designing socio-technical systems for health care, the design of technology is balanced, when done well, by continuous work making sense of the practices enacted by the people and groups who work with and across implemented technologies.

1.2 Motivation, Values, and Interests for Health care

A central consideration in socio-technical design is understanding the underlying logics driving people's work, intended artifacts, and the organizational or institutional cultures involved in any given situation (Mol, 2008). On the one hand, there are the logics of accounting and efficiency motivating a good deal of health care information technology. The administration of health care organizations has responsibility for the financial health of an organization and is typically the sponsor of IT organizations serving health care providers. On the other hand, professional caregivers operate principally in the logics of care, seeking health solutions for people who seek care and that are optimal for them. Institutionally, logics of community building, social capital, institutional capability, and cohesion warrant consideration.

These logics are related to interests, needs, and values of the people and roles that interact within a health care case. With the enlarged ecology of included roles the variety of interests that have to be considered is also extended (Abru Amsha and Lewkowicz, 2017). Unavoidably, there is an intersection of interests (Bossen, 2017). They can support or oppose each other (for example, control and accountability vs. enhancing work life quality). Consequently, flexibility is needed to address the different logics and needs of different roles and people. Additional institutional issues may need to be considered, such legal issues (Abru Amsha and Lewkowicz, 2017), logistics, and environmental safety hazards.

A deliberate analysis of the intertwined interests and values can be approached, for example, by a value network analysis approach (Augl and Stary, 2017). Alternatively, the values and interests can be observed in the formal or informal design rationales discussed during the design of socio-technical systems. Those design rationales are socially constructed (Bossen, 2017) in a networked fashion. Especially, the values and interests of workers in a social environment such as a medical system should inform the distribution of competences and functions between humans and technologies (Bossen, 2017). For example, a typical design question might be: should processes be automated, or should they be overseen and decided upon by humans? At the root of many socio-technical design questions (in health care and otherwise) is a decision concerning for whom the system is designed; this is a question about the values brought to bear on the design process itself.

1.3 Time, Dynamic Constellations, Processes, and Places

A socio-technical perspective recognizes that the set of included people and roles as well as their interests and values is not a stable phenomenon but highly dynamic. Consequently, time and the underlying dynamics are critical dimensions to consider in socio-technical work. For example, Jacobs and Mynatt (2017) focus on health care journeys and choose a long-term view that takes changes over time into account as they, for example, are represented in a series of conversations.

Many cases take the organization of processes as a basis for socio-technical design (e.g., Sarcevic et al., 2017; Augl and Stary, 2017). The design and technical support of processes requires planning and the anticipation of decisions in accordance with standard operating procedures, norms, and regulations. However, planning is hindered by the messiness and unpredictability of concrete constellations. Sarcevic et al. (2017) mention the example of team formation in surgery that varies from patient to patient, and Ackerman et al. (2017) emphasize that care in spinal cord injury cannot be rule based.

The understanding and design of processes are critical decisions but can have high uncertainty. The views from two points in time can be differentiated within socio-technical design. In the relative short term, the workflow and trajectories of a single case can focus on the current lives of patients and their care (as opposed to a focus instead on isolated health problems). On the other hand, when a technology is introduced into a social setting, such as an organization, there can be transitions and metamorphoses, some of which are emergent and not easily foreseen (Orlikowski, 2007). However, an organization's current processes can be analyzed and the trajectory of transitions perhaps influenced. Many of the chapters, most notably Augl and Stary (2017), Prilla and Herrmann (2017), and Bardram and Frost (2017), discuss these issues in depth. This second perspective is important with respect to long-term usage and sustainability.

A socio-technical perspective also acknowledges that medically relevant social interactions occur within and are influenced by a wide spectrum of physical locations and their characteristics. Since the characteristics of a place influence the social relationships and interactions (Sarcevic et al., 2017), design must increasingly examine the physical locations in which a technology may get used. For example, the home has become a place where professional health care can happen and where people grapple with their health on an everyday basis (Abru Amsha and Lewkowicz, 2017). There are many other physical places where health care could be supplemented. For example, Parker et al. (2017) point out that healthy lifestyles could be supported through technology within community centers, schools, churches, and libraries.

1.4 Dealing With Data

With additional technical support of social interactions, the tracking of health-related data is also more and more enhanced—interacting may be recorded, the usage of sensors provides data, and so on. Most of the chapters in this book discuss the possibilities and issues in new forms of data capture.

The chapters in this book show that having awareness of new forms and availabilities of data offers many advantages. For example, the possibilities of knowledge acquisition and revision as well as for lifelong learning are enhanced. However, there are also disadvantages. It requires additional effort on the part of patients, caregivers, and clinicians to pay attention to these data. Clinicians are often concerned with their workflow (Bardram and Frost, 2017), because of time constraints but also if they become responsible for missing data that indicate serious health problems. Additional roles may be needed or involved to handle new forms of available data. Data can also be a mixed blessing for patients and caregivers. Privacy is sometimes invoked to cover the issues of data dissemination, but the problem is more nuanced than either having access or not: for example, if people realize that their actions are no longer invisible this can change their routines.

Socio-technical data design takes into account the type and amount of data that are elicited. This includes questions such as: how do people interact with the data? Is there a difference when focusing on human–data interaction instead of human–computer interaction? Who owns the data and how do data develop? What will be the secondary use of data being generated by the health care system (e.g., to make problems visible) and what will be the consequences of this use with respect to certain social contexts? For example, professionals can be confronted with the results derived from data to make the consequences of their work more transparent, but this may lead to nonadoption or resistance (Bardram and Frost, 2017; Zhou et al., 2017). On the other hand, new forms of care, based on data, may be possible (Abru Amsha and Lewkowicz, 2017).

1.5 Extended Scope of Technology

Along with the expanded scope of social interaction and the increase in data being produced and being relevant, the scope of technology is also changing and growing. Medical technology was once only about medical instruments and appliances, production of images, and control of medical data within hospital or medical systems. It now includes many types of support. Patients (and even people not in immediate need of medical attention) and clinicians all live in a dense ecology of medical technologies. This ecology necessarily includes the obvious range of consumer systems including health applications, mobile technologies, and tablets and other hardware focused on media and information consumption. The ecology for most people includes many other applications, including social networks, other communication systems, forums and webboards, web applications and portals, and even games. The everyday ecology of use for consumers and clinicians alike includes systems to support coordination and collaboration, learning and reflection, and information availability. If we look past the present, we see emergent technologies include sensor-based environments to track and monitor people both for the management of chronic conditions and to watch for potential problems (see Ackerman et al., 2017), robotics both in the home to augment caregiving and in the hospital to support or replace nursing care, and analytical software using Big Data and machine learning to compare among cases and to learn from them.

2. CONSEQUENCES FOR DESIGN

2.1 Focus on Interests and Values

Socio-technical design means taking people's values and interests into account. The term "socio-technical design" has long connoted this consideration: Clegg (2000) emphasized that values and mindsets are central to design, and Mumford (1983) stressed that a design has to fit with individual interests and social values. Given the variety of individuals and roles that are included in the cases of this book, the challenge of taking a wide range of values, interests, and mindsets into account becomes apparent.

A central orientation for socio-technical design in health care is to focus on life goals instead of mere health care goals (Jacobs and Mynatt, 2017), and it is not only about covering situations where people are sick but also supporting ways of living a healthy life between phases of medical treatment. To do this, designs must support the values, needs, practices, attitudes, and interests shared by individuals and families. Such an approach involves studying the intersection of people and their social, organizational, and physical environments. Parker et al. (2017) show this includes trust and social coherence (a sociological construct that describes a group's connectedness). Parker et al. (2017) point out the importance of understanding "... how these relationships inhibit or support health behaviors and attitudes."

Designs must also be aware of the values and mindsets that are socially negotiated and are a subject of continuous change. For example, while it may be important to deemphasize social relations for a person needing extensive care at one point, later having an active, independent social life may become a highly important value (Ackerman et al., 2017).

Finally, the set of the social roles to be included is also the basis for identifying the relevant stakeholders whose interests should be included in the design process (Bossen, 2017; Augl and Stary, 2017).

2.2 Process- and Time-Orientation

Understanding the dynamics of social relationships, interactions, and processes among patients and clinicians is central to socio-technical design. Therefore a design process should include the capturing and the consolidated understanding of the as-is-process as a starting point. Many of the chapters in this book walk through that process.

Within an existing institution or organization, the tasks and the activities during the interaction with patients are often organized and technically supported from a work-flow perspective (Augl and Stary, 2017). Process optimization by digital means is one way to reduce the growing fiscal and social burden of organizing prevention, treatment, and care. Integrated digital workflows can help increase efficiency in health care, in particular through intelligent task allocation and load distribution, through recognizing overhead (e.g., through duplication) of organizational, diagnostic, or therapeutic interventions and through enhancing communication among relevant stakeholders. Gains in efficiency could increase effectiveness, for example, through improving quality of care by maintaining contact with patients between clinical visits. However, from a socio-technical perspective, considering the timeline of a patient's development and of the trajectories between events and situations should not only be a matter of efficiency-oriented workflow improvement but also support a holistic view as a basis for socio-technical design.

The design of so-called patient-facing applications, on the other hand, may involve the design of interactions as much as workflows per se. Maintaining and improving clinical work-flows can be just as important to patient-facing applications, but patients and their caregivers can need more ad-hoc interactions as well (Jacobs and Mynatt, 2017; Ackerman et al., 2017). In fact, for many applications focused on health rather than medical conditions, interaction patterns may be paramount. The games in Parker et al. (2017) are one example. Many involving design for patient-facing applications utilize some version of the so-called human-centered design process, the standard human–computer interaction iterative cycle using requirements gathering, design, and evaluation (Shneiderman et al., 2017).

Design has to also consider that the people, their care networks, their social environments, as well as the clinical settings are continuously facing changes. Many of the chapters involve this problem in one way or another. Bardram and Frost (2017) discuss the changes in a clinic, and Zhou et al. (2017) point out the changes in the use of an electronic health record. Many systems, including those in Bardram and Frost (2017) and in Jacobs and Mynatt (2017) were designed as part of a change process. Ackerman et al. (2017) points to this problem as a major issue with using emerging technologies. Cherns aptly points out "…that the present period of transition is not between past and a future stable state but really between one period of transition and another (Cherns, 1987, p. 159)." This has proven to be especially relevant in the health care domain.

2.3 Complexity and Agility

The dynamics within the health care domain are accompanied by a high degree of complexity characterized as "betwixtness" by Bossen (2017). Betwixtness, in Bossen's view, is caused by the multiplicity of involved stakeholders who represent a huge variety of goals, interests, and perspectives that have to be taken into consideration by socio-technical design and that are, in turn, affected by it. Accordingly, it can be hard to identify all of the "real" stakeholders (for example, other relatives and volunteers in Abru Amsha and Lewkowicz, 2017, and parents in Bardram and Frost, 2017) and to elicit their point of view.

Since socio-technical design should start with a deliberate analysis of the starting situation (which is not trivial) and as it also depends on the perspectives and routines of those who are asked or observed, there are no easy rules that can serve as guidelines to overcome the betwixtness. Several disciplines and methodologies often have to be included in the design process.

To deal with the complexity, an agile approach should be considered for socio-technical design. Often the users and their reactions can only be understood after the effects of first interventions (technical, organizational, or social) can be observed (e.g., in Abru Amsha and Lewkowicz, 2017 or in Bardram and Frost, 2017). More about the problems becomes obvious when the new technology is really used—a deep dive is necessary for gaining deep insights into the problems to overcome (Ackerman et al., 2017). Agility allows customization and reengineering to happen when the design is already in use. However, an agile approach may oppose the robustness that is solicited if not required in health care domains. Customization has to be understood at a level that goes beyond technical infrastructure and includes organizational aspects, roles, tasks, and so on. Flexibility of technical infrastructure has to be complemented by flexibility of those who are in charge.

With respect to managing complexity, it has also to be understood that scalability of a certain design does not work on the social level. For example, many patients want direct contact with medical personnel but this cannot be easily offered at scale by a medical system (Bardram and Frost, 2017).

2.4 Participation

Socio-technical design may be accompanied by a participatory approach that can also be observed in the book's cases. For example, Augl and Stary (2017) promote a stakeholder-centered socio-technical design approach. Of particular interest are collaborative events, such as consultancy or expert panels involving various stakeholders, such as general practitioners, pharmacists, and clinical experts. For each of them, relevant data are to be provided for informed decision-making, in particular for arriving at a valid understanding of the starting point and the ongoing processes. Parker et al. (2017) describe – in the context of public health – community-based participatory research as an approach that focuses on "social, structural, and physical environmental inequities through active involvement of community members, organizational representatives, and researchers in all aspects of the research process." From a practical point of view, a participatory approach has to overcome some obstacles: The driving forces that will make the system work are often not well identified, and participation of the appropriate parties may not be well promoted. This applies especially to informal roles. With respect to the number of potentially includable people, it is hard to systematically decide who should be included (for example, from doctors, nurses, patients, administration, relatives), and who are the problem owners. Furthermore, organizing efficient and effective participation may require that institutional arrangements have to be changed.

3. METHODS THAT ARE APPLIED

How we go about the design of socio-technical systems is another consideration. This section surveys some of the methods used in the cases examined in this book's chapters.

3.1 Empirical Work

The purpose of empirical work is twofold:

- Its results can be used as a starting point to inform the process of design.
- It is needed in the course of a formative evaluation of a design.

There are two types of data that have to be considered: those which are produced in the course of health care and those which a specifically elicited by socio-technical research and design.

The applied methods in the various cases of socio-technical research include:

- Analysis of available health care data.
- Usability studies and clinical workflow studies.
- Ethnographically based methods combining interviews and observations. These are field-based methods, meaning that the designer or researcher engages the situations.
- Focus group sessions, discussion meetings, and design workshops where groups of people are asked questions and/or observed. These tend to be more participatory and are part of the design process itself.
- Simulations (e.g., of trauma situations in Sarcevic et al., 2017).

In the chapters' cases, these methods involved many different kinds of data collection, such as:

- Taking videos and photos in the course of observation to see how actors coordinate their work in situ. This is especially helpful not only in observations, but also in design workshops.
- Document analysis, for example focusing on how notebooks support the collaboration between the care actors. Notes and photos of different liaison notebooks were taken in Abru Amsha and Lewkowicz (2017).

- Temporal analysis. Jacobs and Mynatt (2017) presented journey analysis in their chapter.
- Video review of live events.
- Identification of relevant artifacts. Artifacts used to coordinate and document patients' hospital stays were described in Zhou et al. (2017).
- Using probes to examine how existing tools and resources can support patients' needs (Jacobs and Mynatt, 2017).
- Generating concept maps (Augl and Stary, 2017)

In most of the cases presented in the chapters, various methods were combined over the lifetime of the projects. Of special relevance as a starting point for socio-technical design are observations of changes over time that go beyond the analysis of a certain situation at a fixed point of time. For example, Jacobs and Mynatt (2017) determined how patients' cancer care experiences change over time and were integrated with nonmedical events and challenges. Bardram and Frost (2017) describe the unfolding of a project over a very extended period.

In the course of socio-technical design, empirical methods are relevant not only for the purposes of research, requirements gathering, and initial design but also as a means to provide feedback to the participants and stakeholders who are affected by the design. Empirical examinations and evaluations about the outcome and effects of a specific design should be continuously fed back to the various stakeholders. The evaluations can also be provided as feedback to users as an orientation for their behavior; this is of high relevance for supporting continuous learning and improvement as an integral part of a socio-technical approach. To support this feedback, Sarcevic et al. (2017) emphasize the need for metrics for measuring effects to feed them back to the field.

3.2 Methods to Support Design

The described projects applied various methods to support design, including:

- Mockups and scenarios, sketches, and annotated paper displays to demonstrate the planned solution in an early stage (Prilla and Herrmann, 2017; Ackerman et al., 2017)
- Mockups and other high-fidelity prototypes (Prilla and Herrmann, 2017; Parker et al., 2017)
- Value Network Analysis and subject-oriented business-process modeling which emphasizes the relevance of the included roles and their points of view (Augl and Stary, 2017).
- The Patient-Clinician-Designer Framework (Marcu et al., 2011).
- Facilitated design workshops, where the characteristics of prototypes and the way of using them were inspected by walkthroughs (Prilla and Herrmann, 2017) or co-developed with users (Abru Amsha and Lewkowicz, 2017).

These methods were usually applied in interactive sessions. A typical setting was, for example, a meeting with five members including a physician, a registered nurse, a physiotherapist, and two home helpers (Abru Amsha and Lewkowicz, 2017).

The initial empirical work usually strives to produce substantial amounts of data about the current practices and to support a phase where a great many possibilities and options can be considered. Within the design phase, decisions have to be made and a phase of convergence starts. The formative evaluation of the design outcome can help to produce new ideas but usually pursues the goal of sorting options out to achieve further convergence.

4. CHALLENGES AND PROBLEMS

In addition to raising possibilities—new uses of technology, new organizational forms, new types of health-related interactions, and re-designs of social processes—the chapters raised a number of challenges and problems that can be best seen from a socio-technical perspective.

4.1 Dealing With Health care Data

Above we discussed some of the issues with health care data. Electronic support of health care helps to increase the available data about patients, clinical workflows, administrative procedures, and so on. These data are expected to be beneficial for managing health care and supporting patients. However, even if a considerable amount of data are produced, it is also possible that a substantial amount of relevant data might remain unconsidered (Abru Amsha and Lewkowicz, 2017) or not evaluated by clinicians and caregivers (Ackerman et al., 2017; Zhou et al., 2017). Furthermore, people require trustfulness; the more data are captured the more privacy seems relevant and important (Abru Amsha and Lewkowicz, 2017). With respect to data elicitation, it is an open question who is allowed to provide data, especially when the social environment of patients is taken into account. In the MONARCA case (in Bardram and Frost, 2017), it appears questionable whether parents should be allowed to enter data about the behavior or mood of a patient.

As the amount of data becomes more extensive, this can cause problems with respect to ongoing usage. The belief that clinicians can or should be continuously monitoring patients turns out to be overly naive. Clinicians usually see a large number of patients, have limited amounts of time, and therefore have to make prioritization decisions about what data are important. They may as well need to delegate the oversight of the data. In the case of Bardram and Frost (2017), a patient's data was monitored by a trained nurse.

Although the amount of available data might be larger, it is not necessarily complete. A fragmentation of patient records and data can be observed in our chapters (Bossen, 2017; Zhou et al., 2017). Often, only single episodes within longer health problems are stored but the whole situation and history of a patient can be difficult to determine. This is especially inappropriate when patients have comorbidities or outstanding psychosocial issues (Zhou et al., 2017). Recording entire trajectories of a patient's situation is not regularly done.

Striving for a complete data set has also disadvantages. Clinicians may get lost in details instead of building a holistic overview or may have to invest too much work in evaluating the data. Furthermore, there are many more possibilities for privacy problems as the data set about each person becomes more extensive. Health care institutions are aware of the privacy impacts of their data, and they are afraid that recorded data might document problems about dealing with patients, causing potential liability issues (Prilla and Herrmann, 2017).

4.2 Trust

The chapters discuss a series of problems with trust that affect the success of socio-technical design if not sufficiently taken into account: people (both clinicians and patients) may not trust the developers will have gotten the design right and hence may expect that the technical system will not meet their goals. Patients may not believe that the system will maintain their privacy and that the confidentiality of personal data will be guaranteed. This mistrust includes a concern that unauthorized people can look at their personal data.

This mistrust is related to a lack of transparency of ongoing operations and about the group of people who are allowed to see certain data. The extension of the ecology of included roles implies that more and more people may be involved in a medical situation to whom certain personal data should not be disclosed. This may also apply to family members (Sarcevic et al., 2017).

4.3 Complexity and Limited Perspectives

The lack of transparency is related to the complex social arrangements and interactions in the health care domain. The betwixtness, lack of structures, and lack of rules make it difficult for the involved people to understand and to anticipate what is going on. This is especially a problem for a participatory design process since the participants may be unable to correctly understand the relevant issues and anticipate how a new technology will comply with their needs and values.

A naive way of reacting to this complexity is, from the viewpoints of the involved actors, to orient themselves to their superiors within an organizational hierarchy or to accept simplified linear models of cycles, workflows and processes. For example, most dominant design viewpoints neglect the iterative, cooperative process of finding, filtering, valuing, and connecting information that is involved when clinicians reason about patient cases (Bossen, 2017).

Instead, successful socio-technical designs take into account the real complexity of unstructured and spontaneous processes instead of arguing for rule-based workflows. Simplified views about processes and collaborative interaction may result in the neglecting of valuable information and in a misrepresentation of the complexity of health care reality. Furthermore, the socio-technical support of information acquisition and information exchange should avoid a dominance of hierarchy-driven information exchange and allow for bottom-up information elicitation and distribution (as in Abru Amsha and Lewkowicz, 2017).

4.4 Lack of Willingness

In several of our chapters, a lack of willingness to use a new application or technology could be observed. In others, the intended users did not integrate the current tasks and the new technology very well. Prilla and Herrmann (2017) found especially staff showed a low willingness to use new technology for communication purposes, and Bardram and Frost (2017) found that the communication support tools were used in other ways than originally intended.

Users and decision-makers always compare the new possibilities of a socio-technical design with established routines or with legacy tools and with the organizational costs of the change. There can also be a decline over time: people may use the system at the beginning but not on an on-going basis. Participants often expect that no additional costs will arise, for example when patients get the opportunity to report frequently on their condition (Bossen, 2017). For professionals, the need for entering of extra data can cause resistance—therefore smoothly integrating data capturing into the routine tasks is important for success (Sarcevic et al., 2017; Prilla and Herrmann, 2017). The process of adoption can be incremental or disruptive.

In many cases, the reason for resistance against using a new technology—or the reasons why people would use a technology—are not well understood. Indeed, if there is some resistance it may not be clear what influence it will have on the change process overall. For example, sometimes medical staff may argue that newly introduced socio-technical processes do not comply with their established routines and that therefore there is no time available to use the new technology. However, Prilla and Herrmann (2017) found that after a while there are people who use the new functionality without problems—although some people still argued that it was not possible. Thus, there is a difference between what stakeholders think they want and what they really want. It is a challenge to sufficiently present the stakeholders' goals in the socio-technical process and, as Prilla and Herrmann (2017) also showed, to continuously make them a subject of reflection.

People have various reasons for adopting new technology. It could be based on evidence or approval by others; people judge systems subjectively with respect to the perceived achievement of their goals. Some adoption is based on opportunistic behavior or deliberative reflection. It is advantageous if networks of adoption arise, but the emergence of those networks does not guarantee success. However, we know that having advocates in the environment of a user will influence whether adoption takes place. Somebody should be responsible for introducing the new system, but these roles or tasks are in many cases not clearly specified. The question is which roles make the system work. The driving forces are often not well identified and promoted, especially the roles of intermediaries in an organization's communication, and this argues in favor of an iterative process of adoption and adaptation (Jacobs and Mynatt, 2017; Sarcevic et al., 2017).

4.5 The Evidence Problem

One central issue throughout many socio-technical design projects is how to provide evidence that the pursued approach will produce a benefit for the involved stakeholders from their point of view. There is a lack of agreed-upon metrics for measuring a system's effectiveness (Sarcevic et al., 2017). It is even hard to identify the relevant indicators that mirror successful initiation or completion of change. The strength or shortcomings of a socio-technical design becomes most apparent under the conditions of crisis—however, these conditions cannot be systematically reproduced during the evaluations to bring evidence for the utility of a new technology and its accompanying organizational (or clinical) change.

The problems with evidence point toward two types of vicious cycles: On the one hand, organizations must be willing to support changes—for example, of the available infrastructure (Prilla and Herrmann, 2017)—so that the new technology can be effectively introduced and used. Otherwise, the benefit of the socio-technical design cannot be demonstrated. However, organizations are not willing to make effective changes before they have evidence that it is worth doing so. On the other hand, the individuals and stakeholders must be willing to take part in evaluations, which usually cause extra effort in the beginning. Again the problem is that users and management will accept any additional workload only if they are convinced that this will lead to a benefit. That means they need evidence before they will become part of the evaluation.

5. HOW TO DEAL WITH THE PROBLEMS

5.1 How to Increase Motivation

Increasing the willingness of organizations and people to take part in a socio-technical project, especially in the early phases of formative evaluation, is an essential success factor. The evaluation phases are of crucial relevance since the underlying complexity often does not allow for sufficient anticipation of the effects that will be produced by the socio-technical design. Health care brings an additional critical constraint—it has to be accepted that every intervention must not disturb the reliability of clinical routines.

To increase willingness, one way is to have the introduction of new technology continuously reflect its impact on the values and interests of the participants, for example, with respect to the payment structure. The design of socio-technical health care processes should include continuous informing of stakeholders about the status and quality of health care processes in general and the risks and side-effects in particular. Effective sharing of information by articulation and documentation can take place middle-out (Augl and Stary, 2017) instead of primarily top down or bottom up. An incremental transition from divergent information eliciting to producing convergence is required: "Stakeholders may begin from different points when reflecting the situation as-it-is while heading in the same direction when mindfully creating and publically committing to proposals for change (Augl and Stary, 2017)." The promoters of a project must be continuously aware of how an intervention changes the stakeholders' points of view.

It is likely that willingness to use a new technology will be higher if it amplifies people's existing practices (Jacobs and Mynatt, 2017) instead of introducing new tasks that firstly have to be integrated into daily routines (Prilla and Herrmann, 2017). However, new socio-technical designs must also consider that existing routines are not stable over time and users' needs will repeatedly change.

It is most reasonable to consider the system of social interactions between the potential users and relevant stakeholders as a basis to build increased motivation and willingness to participate. One way to do this is to organize events for initial trust building between people where they can realize that the involved actors are striving to support each other (Abru Amsha and Lewkowicz, 2017). Using community-based arrangements may be another option (Parker et al., 2017). Further support results from identifying potential lead users who might serve as role models and as early adopters of new technologies and organizational procedures. Such a behavior modeling for accepting socio-technical interventions could also be achieved by visiting, contacting, or observing other organizations or communities where role models are already active.

5.2 Improving the Quality of Data Handling

The way of handling the data that are produced and elicited during health care processes has been a central issue. It greatly influences the stakeholders' willingness to participate. Ensuring data integrity and security influences users' willingness to provide confidential data. Trust issues about and the transparency of what occurs and which effects are produced can be central factors of acceptance with respect to data handling. In particular, as health care data are highly sensitive data; when digital patient record management is introduced or extended especially in patient-facing applications, the management of that data needs to be transparent for patients and other users. The availability of data is highly related to trust issues. On the one hand, more availability and sharing leads to others having a better understanding how to act and decide. On the other hand, availability and sharing can violate confidentiality and legal restrictions. The tension can be seen in Zhou et al., 2017, with concerns about making psychosocial statements about patients in the official documentation. This contradiction cannot be generally solved but requires a deliberate analysis in specific cases with a fair reflection of the interests and mindsets of the involved stakeholders.

A further issue is the quality of data. Data should include not only mere measures or "facts" but also explanations. Challenges should be highlighted and trajectories made visible. All in all, discussion-oriented documentation should be available in many situations, especially those concerning complex chronic conditions (Abru Amsha and Lewkowicz, 2017; Zhou et al., 2017).

When patients or their social environment start to upload data about their situation by themselves, privacy concerns become obvious. Designers of health tools must consider how a new technological system provides or hinders users' control over their data. Again, those conflicts can only be solved with respect to the specific conditions of each case. Furthermore, introducing certain organizational rules for privacy is not sufficient but has to be completed by guidelines and training, an understanding of regulations, and an understanding of emergent conflicts (Prilla and Herrmann, 2017).

5.3 Control and Flexibility

One way of dealing with a low level of foreseeability and a high level of betwixtness is to offer control to the stakeholders and a sufficient degree of flexibility. In a socio-technical approach, flexibility and possibilities for customization offer stakeholders the possibility to participate in what is often called design-in-use, and to adapt the design to their needs.

One approach is to offer under design; that is, to specify only those constraints that are legally and technically unavoidable and to leave sufficient room for users to react to emerging events and conditions (Fischer and Herrmann, 2011). Within socio-technical design, flexibility and customization do not only refer to technical functions but also to plans, organizational structures, workflows, etc. For example, care plans have to be continuously tailored (Ackerman et al., 2017).

To allow for customization not only increases control and increases use, but it also is a means to make the interests and values of the participants visible to the designers and researchers (Jacobs and Mynatt, 2017). How participants customize mirrors their needs and expectations. In health care, of course, there is a tension between the degree of possible flexibility and the required reliability.

5.4 Facilitation and Improved Quality of Communication

Successful communication is one of the key aspects of socio-technical design. On the one hand, communication is the basis and driving force of the design process. Creating shared understanding must be put into the foreground. Different viewpoints and perspectives have to be brought together without a preference for hierarchy or formal positions of the stakehold-ers (Augl and Stary, 2017). To achieve this integration, facilitators or mediators may need to be involved. Certain methods like walkthroughs (Prilla and Herrmann, 2017), continuous and immediate note taking, and visualization of people's contributions can be helpful. Superiors have to be trained to control their influence on the process if the beliefs and perspectives of other stakeholders are to be accumulated and used.

On the other hand, technical and organizational support for communication itself is a topic of socio-technical design. For example, informal communications and social media can play an increasingly important role alongside formal documents. This is especially true for patient-facing applications like those found in Jacobs and Mynatt (2017); their system's goal was par-tially to facilitate informal communication between clinician and patient. Coordination tools, like those found in Abru Amsha and Lewkowicz (2017) to facilitate ad-hoc communications among care providers, are also important. The formats and tools for communication sup-port have to be in compliance with communication literacy of the participants, including the medical context, patient-understandable language, and e-health care netiquette.

Aside the information design to support existing patient–physician relationships, direct interaction channels and shared repositories need to be provided for continuous care taking and case management. Stakeholders need to know who is going to set the next activity and who is currently taking care about the case or patient. Thereby, response and delivery times play a crucial role.

5.5 Providing Evidence

Evidence helps with the assessment of socio-technical design and provides a basis for mak-ing benefits apparent and increasing the stakeholders' willingness to participate. Within the course of socio-technical design, decision-makers may request evidence for the benefit of the system. Bardram and Frost (2017) follow the HIMSS STEPS model, referring to increased sat-isfaction (S), improved treatment (T), increased electronic data quality (E), increasing preven-tion (P), and/or economical savings (S). However, effective metrics to measure any increases related to these aspects must be further developed (Sarcevic et al., 2017).

Alternatively, evidence can also be achieved by subjectively convincing participants of certain effects. Incremental steps that demonstrate the benefits of certain socio-technical efforts with respect to well-known tasks can help to convince people of the advantages of a socio-technical design. Therefore, demonstrations and prototypes can help to make the technical functionality understandable, increase further motivation, and make benefits can become apparent.

As a socio-technical design progresses, there is the possibility to discover that the domi-nant logics incorporated into a design are not aligned with the values of the people who will use the system. In such cases, an iterative design process can uncover and then incorporate the multiple logics and perspectives of diverse stakeholders.

6. SUMMARY AND FUTURE WORK

In summary, the field of socio-technical design in the context of health IT systems is sub-stantially further than it was 20 years ago. We have a better understanding of the issues involved in the design and use of technical systems within social systems. We also have a better understanding of the issues involved in changing social systems, especially within organizations, through technologies. The book chapters and this conclusion highlight many of those issues. Additionally, we have a far better understanding of how to meaningfully incorporate the needs and values of patients, clinicians, and other stakeholders in the design and usability of health IT systems. We have learned to design using rapid, iterative design cycles with multiple formative evaluations. The book chapters and this conclusion also highlight many of those methods.

Ultimately, however, the design approaches, techniques, and outcomes discussed in the chapters are not complete. Many of the chapters ended not only with findings but also with questions. Indeed, there are still many open questions in this conclusion.

Looking forward, we see an even more vital requirement to incorporate a socio-technical viewpoint into the design of health IT systems of all types. We are on the verge of many new technologies becoming widespread—not only mobile technologies but also next-generation computational environments that include machine learning, multi-modal interfaces, and ubiquitous sensors. Bio-sensors and data will become increasingly pervasive and require novel infrastructures including complex event processing. Designing suitable health IT systems using new technologies will be a constant challenge.

We also expect new theories to emerge that will forward socio-technical understandings. As this conclusion and many chapters have discussed, researchers and practitioners alike are grappling with the need to understand how the details of the technologies in their use and the details of people's activities (or practices) interact and actually come to mutually constitute each other (Orlikowski, 2007). As health and medical IT systems move from the hospital and medical institutions into people's homes and everyday lives, these new theories will be of great use.

The challenges in health care are daunting. We fully expect the needs and changes in medicine to continue if not accelerate, and as well, technological innovation of all sorts will continue if not accelerate. We firmly believe that the lessons in this book will be indispensable.

References

Ackerman, M.S., 2000. The intellectual challenge of CSCW: the gap between social requirements and technical feasibility. Human-Computer Interaction 15 (2–3), 179–204.

Ackerman, M.S., Buyuktur, A.G., Hung, P.-Y., Meade, M., Newman, M.W., 2017. Sociotechnical design for the care of people with spinal cord injuries. In: Ackerman, et al. (Ed.), Designing Healthcare that Works. Elsevier, Amsterdam.

Abru Amsha, K., Lewkowicz, M., 2017. Supporting collaboration to preserve the quality of life of patients at home – a design case study. In: Ackerman, et al. (Ed.), Designing Healthcare that Works. Elsevier, Amsterdam.

Augl, M., Stary, C., 2017. Adjusting capabilities rather than needs in computer-supported daily workforce planning. In: Ackerman, et al. (Ed.), Designing Healthcare that Works. Elsevier, Amsterdam.

Bardram, J.E., Frost, M.M., 2017. Double-loop health technology: enabling socio-technical design of personal health technology in clinical practice. In: Ackerman, et al. (Ed.), Designing Healthcare that Works. Elsevier, Amsterdam.

Bossen, C., 2017. Socio-technical betwixtness: design rationales for healthcare it. In: Ackerman, et al. (Ed.), Designing Healthcare that Works. Elsevier, Amsterdam.

Cetina, K.K., Schatzki, T.R., Savigny, E.von (Eds.), 2005. The Practice Turn in Contemporary Theory. Routledge, London.

Cherns, A., 1987. Principles of sociotechnical design revisited. Human Relations 40 (3), 153–161.

Clegg, C.W., 2000. Sociotechnical principles for system design. Applied Ergonomics 31 (5), 463–477.

Fischer, G., Herrmann, T., 2011. Socio-technical systems: a meta-design perspective. International Journal of Sociotechnology and Knowledge Development 3, 1–33.

Jacobs, M., Mynatt, E.D., 2017. Design principles for supporting patient-centered journeys. In: Ackerman, et al. (Ed.), Designing Healthcare that Works. Elsevier, Amsterdam.

Jahnke, I., Ritterskamp, C., Herrmann, T., 2005. Sociotechnical roles for sociotechnical systems – a perspective from social and computer sciences. In: AAAI Fall Symposium Proceedings, vol. 8, .

Kaziunas, E., Ackerman, M.S., Veinot, T.C.E., 2013. Localizing chronic disease management: information work and health translations. Proceedings of the American Society for Information Science and Technology 50 (1), 1–10.

Kling, R. (Ed.), 1996. Computerization and Controversy: Value Conflicts and Social Choices. Morgan Kaufmann, San Francisco.

Marcu, G., Bardram, J.E., Gabrielli, S., 2011. A framework for overcoming challenges in designing persuasive monitoring systems for mental illness. In: Proceedings of Pervasive Health 2011. IEEE Press, pp. 1–10.

Mol, A., 2008. The Logic of Care: Health and the Problem of Patient Choice. Routledge, New York.

Mumford, E., 1983. Designing Human Systems for New Technology: the ETHICS Method. Manchester Business School, Manchester, UK.

Orlikowski, W.J., 2007. Sociomaterial practices: exploring technology at work. Organization Studies 28 (9), 1435–1448.

Parker, A.G., Saksono, H., Hoffman, J., Castaneda-Sceppa, C., 2017. A community health orientation for wellness technology design and delivery. In: Ackerman, et al. (Ed.), Designing Healthcare that Works. Elsevier, Amsterdam.

Prilla, M., Herrmann, T., 2017. Challenges for socio-technical design in healthcare: lessons learned from designing reflection support. In: Ackerman, et al. (Ed.), Designing Healthcare that Works. Elsevier, Amsterdam.

Sarcevic, A., Marsic, I., Burd, R.S., 2017. Dashboard design for improved team situation awareness in time-critical medical work: challenges and lessons learned. In: Ackerman, et al. (Ed.), Designing Healthcare that Works. Elsevier, Amsterdam.

Schroeder, R., 2007. Rethinking Science, Technology, and Social Change. Stanford University Press, Palo Alto.

Shneiderman, B., Plaisant, C., Cohen, M., Jacobs, S., 2017. Designing the User Interface: Strategies for Effective Human-Computer Interaction. Pearson Education, Upper Saddle River, NJ.

Wulf, V., Rohde, M., Pipek, V., Stevens, G., 2011. Engaging with practices: design case studies as a research framework in CSCW. In: Proceedings of the ACM Conference on Computer Supported Cooperative Work. ACM, New York, pp. 505–512.

Zhou, X., Ackerman, M.S., Zheng, K., 2017. The recording and reuse of psychosocial information in care. In: Ackerman, et al. (Ed.), Designing Healthcare that Works. Elsevier, Amsterdam.

Index

Printed in the United States
By Bookmasters